Reflecting on the Future of Academic and Public Libraries

Reflecting on the Future of Academic and Public Libraries

Editor

Roger Burnett

scitus academics

Reflecting on the Future of Academic and Public Libraries
Edited by **Roger Burnett**

ISBN: 978-1-68117-636-9
Library of Congress Control Number: 2016946496

SCITUS Academics LLC,
616, Corporate Way, Suite 2, 4766,
Valley Cottage, NY 10989

www.scitusacademics.com

Notice

Printed in United States of America on Acid Free Paper

Preface

Academic and public libraries are now very different from 20 years ago. In addition to public libraries, academics have to face constant changes these days. In addition to technological disruption in the media, scholarship and education, they face a gradually harsh budgetary environment as libraries modernize. Uncertainty and transformation are part of the political, social and technological environment. College and faculty libraries are at the heart of the library system changes. Academic libraries have to serve groups of users with a variety of research and information needs. In addition to the structural challenges faced by all libraries, academic libraries must also face a rapidly changing educational and publishing environment in which the value and cost of a faculty education is now being called into question. More accountability means that academic libraries must demonstrate their value and also the importance of the teaching system. The growing need for publishing services suggests new roles and responsibilities for academic libraries in future. Each institution needs a unique vision for the construction of an intellectual center for teaching, cooperation and learning in the 21st century. In addition, this vision must reflect the very own voice of the institution and ask what it means to support the development of academic bursaries. In order to foster learning and discovery within a powerful academic framework, libraries must now foster a good ecology of relationships, connectivity settings and tools. But public libraries are currently at a turning point. The way we access and consume information has changed dramatically in the 21st century, giving public library systems worldwide significant opportunities and challenges. Several of our reading habits were changed with the introduction of new technologies. But our need for shared, community-centered spaces to find information and links with others will probably not change in the near future. In order to survive and remain relevant in the electronic age, public libraries must be innovative and courageous. They should embrace the physical and virtual.

This book provides an overview of the leadership skills required to succeed in deliberately changing libraries on a large scale. The book should be a great guide for library managers, academics and public librarians, students of LIS and academics and anyone interested in the future of libraries.

TABLE OF CONTENTS

1

CHAPTER

The Handheld Library: Developments at the Rector Gabriel Ferraté Library, UPC

Beatriz Benítez Juan, Javier Clavero Campos, Miquel Codina Vila, Andrés Pérez Gálvez

Universitat Politècnica de Catalunya (UPC-Barcelona Tech), Barcelona, Spain,

ABSTRACT

The purpose of this paper is to highlight the mobile services developed by the Rector Gabriel Ferraté Library (BRGF) of the Universitat Politècnica de Catalunya (UPC-Barcelona Tech) in Barcelona, Spain.

We hope this paper will be of use to other libraries exploring new technologies for communicating and delivering their services to users at a time when mobile services are an emerging topic in librarianship and information science literature. By setting out the successive steps involved in the as yet unfinished process of building our mobile services portfolio, we aim to offer a detailed picture of the mobile services and features offered by a university library from a case study perspective.

The main topics to be discussed include:

- The BRGF's mobile website, including the information available, its interactive capabilities and the services it provides to its users.
- The mobile-friendly version of UPCommons (the UPC Library Service's institutional repositories).
- The UPC Library Service's mobile OPAC.
- The mobile version of u-win (BRGF's videogame service).
- The use of QR codes to deliver information to mobile devices.
- Text message notifications.

Additional topics for discussion include:

- The library's organisation and the organisational concepts that underpin and make possible its technological developments (including mobile).
- BRGF's concern regarding the reduction of investment in the development of mobile services.
- The criteria and tools used to guide the library's decisions regarding the design and orientation of current and future mobile services.
- How mobile services can help to improve the image of the library as a leading technology site.
- Selected mobile features that BRGF plans to offer in the near future.

Ultimately, this paper aims to delineate the effectiveness and potential of delivering library services by the preferred means of communication of a new generation of students and teachers.

Keywords: academic libraries; mobile services; mobile libraries; mobile communication systems; mobile devices; handheld devices

INTRODUCTION

The Rector Gabriel Ferraté Library (BRGF)[1] is located on the North Campus of the Universitat Politècnica de Catalunya (UPC-Barcelona Tech)[2] in Barcelona. The BRGF is part of the UPC's Library and Documentation Service (SBDUPC) and the largest library at the UPC,[3] accounting for approximately 25% of the total volume of the university's library system.

The BRGF offers direct services to nearly 8,000 students and faculty members from the Barcelona School of Informatics (FIB),[4] the Barcelona School of Telecommunications Engineering (ETSETB),[5] the Barcelona School of Civil Engineering (ETSECCPB),[6] 22 research departments and their administrative staff and services, and many of the university's general services (Human Resources Service, Language and Terminology Service, International Relations Service, Accounts and Finance Service, etc.).

The disciplines taught on UPC's North Campus foster highly technological environment in which students and faculty members are not only quite skilled in the use of information and communication technology (ICT), but also use it regularly in both their academic and personal lives. It is a place where the latest proposals and innovations in the sector to come on the market quickly find users and consumers.

In this environment, the BRGF is emphasising the use of new technologies and digital tools to carry out new projects and improve existing services. This push is in keeping with its commitment to working on innovative services and projects that both cater for the needs of students and faculty members and help to make technology the library's hallmark.

From this perspective, the UPC libraries believe that mobile applications which not only facilitate access to information resources from anywhere and by means of any mobile device, but also take advantage of the capacity of these devices to offer users access to library services, will be (indeed, already are) critical to bringing our products and services closer to users.

In short, we are convinced that accompanying users at all times by means of their mobile phones will soon be more of a necessity than an option. Consequently, one of our technological priorities is to develop the necessary services. To this end, BRGF decided some time ago to offer a series of mobile applications, even though the time was not yet entirely ripe for immediate success. In the current highly competitive informational context, BRGF believed that the greatest risk was to take no risks at all.

This paper includes the following:

- Some reflections on the technological context of academic libraries.
- The main reasons why BRGF and SBDUPC chose to develop the aforementioned applications and an explanation of how they were ultimately developed.
- A list and brief description of the main mobile services developed by BRGF and SBDUPC.
- Proposals for future mobile services at BRGF and SBDUPC.

- We hope that sharing our experience will contribute to raising awareness among academic libraries of the new possibilities for developing mobile services in their respective environments and encourage librarians to approach users in a way that allows them to meet student and faculty service needs more effectively.

CONTEXT

It is clear at a glance that mobile services are a fully integrated part of university students' lives. Mobile technologies, and especially mobile phones, are part of their everyday gear, and mobile phones are the electronic device of choice for today's mobile generation (see data from Ofcom, 2011 below).

More generally, at this point, few would debate the pre-eminent position held by mobile devices in technologically advanced societies. Their ubiquity notwithstanding, we believe it is worth offering some data, primarily from Spain and Europe, that may help to quantify the issue and contextualise the importance of taking these communication devices into account when designing and implementing library services:

- In Spain, there are far more mobile telephone lines than inhabitants: specifically, as of January 2010, there were 52,810,480 lines, according to data published by the Spanish Telecommunications Market Commission (Comisión del Mercado de las Telecomunicaciones, 2010), but just 47,021,031 inhabitants, according to data from the Spanish National Institute of Statistics (Instituto Nacional de Estadística, 2010).
- According to Nielsen (2010):

- In Spain, 21% of young people between the ages of 15 and 24 use their mobile phones to access the Internet. In fact, this was the most popular mobile phone feature among all those examined in the report (the figure rises to 46% in the United Kingdom and stands at 24% in Italy and 20% in Germany).

- Some 38% of mobile phones belonging to Spaniards between the ages of 15 and 24 are smartphones. This figure is 47% in Italy, 36% in the United Kingdom, and 29% in Germany.

- Some 58% of Spaniards between the ages of 15 and 24 are advanced users, using their mobile phones for far more than just voice calls and text messages. This figure is 68% for the United Kingdom, 62% in Italy and 47% in Germany.

- According to Ofcom (2011), in The Communications Market 2010 (August), using mobile phones is the media activity that Britons between the ages of 16 and 24 would miss the most, namely: 33% compared to 28% who would miss television the most and 21% who would miss the Internet the most. It hardly seems risky to venture that these figures will be similar for young Catalans.

- Fundación Teléfonica (2010) also offers a series of figures worth bearing in mind when positioning Catalan and Spanish libraries with regard to mobile services:

  - Out of all homes with Internet connections, 25% connect by mobile phone and 13.8% by video-game console.

  - 3 out of ever 10 Spanish citizens already own 3G-enabled devices, and 1 in 10 has a smartphone.

  - Smartphone sales across Spain increased by 90% last year.

  - The percentage of smartphone users who connect to the Internet is 15 times higher than that of feature-phone users.

  - Some 86% of Spanish citizens believe that using the Internet facilitates learning.

- According to the Federation of Spanish Publisher Guilds (Federación de Gremios de Editores de España, 2010), 5.3% of the Spanish population already regularly reads e-books. Of this group, 6.9% use mobile devices other than e-readers to do so, while only 1% use e-readers.

It is worth noting that the UPC libraries are convinced that these figures are significantly higher among the university's student body and faculty and that the expansion of mobile features into fields such as entertainment, administrative and business services and formalities, shopping, interpersonal communication, etc., together with the increasing implementation of technologies associated with mobile devices, such as Bluetooth, QR codes, global positioning systems or augmented reality, suggest that the trend towards growing use of and demand for mobile features will become even more dramatic in our environment.

TO APP OR NOT TO APP

As some of the above figures suggest, the growth of smartphones has been quite sudden; just a few short years ago, these types of telephones were not at all common among mobile phone consumers in Catalonia. However, since the first tactile devices appeared, smartphones have met with unmitigated success. The emergence of phones such as the iPhone or Android marked a before and after in mobile telephony and created a new market with its own business model based on online application shops (appmarkets) and a strong trend towards the development of software specifically suited for mobile devices, i.e. the aforementioned, and by now omni present, apps. These apps have met with unqualified success and are being used more and more intensively. So why did BRGF decide to develop a mobile website instead?

First, when we began to develop and offer our mobile services, in the first half of 2009, apps were hardly common place among our users. In fact, as we will continue to stress throughout this paper, even ordinary smartphones were not particularly widespread, let alone smartphones with tactile screens. It thus seemed too risky to base our initial offer on apps and simply hope that they would have at least a minimal impact on our users.

Second, we did not wish to be bound to a single platform. At present, for an application to be compatible with the main mobile phone operating systems — iOS, Android and Symbian — it must be manufactured three times, as each system has its own development environment. Given the UPC libraries' limited resources, universal compatibility was thus unthinkable: to develop an app, we would have to have chosen a single platform, thereby excluding all users who used a different one. We believed that, even though it meant sacrificing some of the benefits of an app, by choosing to develop a mobile website instead, we would be able to reach more users, as all they would require was a mid- or high-end mobile with a browser and an Internet connection. Ultimately, we gave more importance to achieving maximum dissemination than to having the sleeker, more powerful type of interface possible with an app.

Third, the library already had staff members with ample experience in web design. This meant that developing a mobile website would entail virtually no learning costs, as they would only need to learn a few key differences between normal and mobile web design. In contrast, developing an app would have required training in and mastery of the development tools and platforms for the chosen system, with the ensuing commitments in terms of time and cost. Likewise, given that it would have meant working in a new development environment, it would surely have affected time it would take ultimately to roll out the service.

In other words, creating a service platform based on a mobile website had a very low cost in terms of time and resources and ensured a swift roll-out, both of which were obviously key factors leading to the choice for this option.[7]

In any event, although the UPC libraries remain committed to the website format as their primary option, above all because they consider it to be the best choice to maximise impact, the future development of a supplementary app for one or more of the main platforms on the market has not been ruled out, with a view to capitalising on the benefits that such a solution might offer for the current system.

BRGF'S MOBILE SERVICE PORTFOLIO

BRGF's Mobile Website *[8]*

One of the most widespread conclusions in the literature on mobile library websites is that they should not be merely a literal translation of the traditional website. Instead, they should be a simplified version offering content and services that have truly been adapted for mobile use.

BRGF's mobile website was designed according to these criteria and is intended to be accessed by mid- to high-end mobile devices. BRGF decided to target these devices in order to offer a more attractive homepage than it could have, if it had catered to lower-end devices, which tend to allow only text and hyperlinks.

Other principles taken into account included:

- In addition to offering information, the site had to include service request forms and enable interaction with end users.
- All retrieved content had to be original; under no circumstances would replicas be allowed.
- Access to content had to be real-time, never delayed.
- The Smubtool would be used to ensure dissemination in social networks by offering a wide range of options for sharing our products and services with a simple click. This tool has also been used for the mobile versions of UPCommons, the UPC library system catalogue and u-win.

Based on these initial parameters, BRGF's mobile website was fully developed by the library's technical staff using the programming languages HTML, JavaScript and PHP and the database manager MySQL. Initially, consideration was given to the option of using an automatic converter — an application that is generally run on a website and automatically generates a mobile version thereof. However, following numerous tests with different converters, this option was ruled out because, due to the structure and design of the website, the results were unsatisfactory: the site's structure became unclear, the images were knocked out of place, text was cut off, etc., making the final website confusing and difficult for end users to browse.

Usage figures for BRGF mobile services are presented in Table 1.

Table 1. Usage figures for BRGF mobile services

Usage figures for some of BRGF's mobile services April 2010 – April 2011			
Service	**Visits**	**Page views**	**Pages/ Visit**
BRGF website	2,135	8,419	3.94
Ooohweb!	110	378	3.44
u-win	261	851	3.26
OPAC	1,879	6,497	3.46
UP*Commons*: E-prints	198	491	2.48
UP*Commons*: Master's theses	308	766	2.49
UP*Commons*: Journals	80	184	2.3
UP*Commons*: Video library	82	266	3.24

At present, approximately two years after the site was launched, the features available from BRGF's mobile homepage (see Figure 1) can be grouped into:

- Major sections also commonly found on the traditional websites of academic libraries: news, services, collections, information about the library, contact details and help.
- Less common applications, some of which are nevertheless among the most popular in the mobile environment, such as information about room and equipment availability:
    - Equipment:

Figure 1. Mobile version of the BRGF homepage.

        - Availability of laptops for loan: The UPC libraries' laptop-lending service has become one of its most popular services. Users can use the mobile website to check the availability of laptops at the library and to see when the next ones are scheduled to be returned. Moreover, if all laptops are currently checked out, users can use a link at the foot of the page to call the library directly from their mobile phone and add their name to a waiting list so as to be notified by text message as soon as the next one becomes available.
        - Availability of desktop computers in the library's ClickArea, where the public desktop computers are located.
        - Availability of e-book readers, USB memory sticks, calculators and video-game consoles available on loan from the library.
    - Rooms: Information on the terms of use, availability and reservation of the group study rooms offered by the library.
    - La BRGFpregunta [BRGF wants to know]: Participation in the polls conducted by the library to determine users' opinions on a specific topic at a given time. The polls are posted in a prominent position on the library's website. Answers are not binding, but they are often used to help orient the services. Questions are usually closed-ended, although users can sometimes choose more than one answer (see Table 2).
- Access to UPC's open-access institutional repositories (UPCommons), the UPC library system's mobile catalogue and the mobile version of u-win, BRGF's video-game service area. All of these tools will be discussed in further detail below.

Table 2. Results of BRGF polls on mobile issues

Polls conducted at the BRGF on mobile devices and services
1. What information or services would you like to be able to access through BRGF's mobile website? [Survey date: July 2009] [Most popular answers: Group study room reservation, 31%; Opening hours: 21.7%; Laptop reservation: 16.7%]
2. How long do you spend browsing the Web with a mobile device? [Survey date: February-March 2010] [Answers: I do not use my mobile to go online, 47.6%; More than one hour a day, 29.4%; Less than one hour a day, 16.3%; More than one hour a week, 6.7%]
3. In addition to sending text messages and making calls, what else do you use your mobile phone for? [Survey date: September-October 2010] [Most popular answers: As an alarm clock, 14.9%; To check the time, 13.5%; To take pictures, 9.3%]
4. What type of mobile telephone do you use? [Survey date: January-February 2011] [Answers: Multi-touch smartphone, 42.7%; feature phone, 34.2%; non-multi-touch smartphone, 23%]
5. Are you familiar with QR technology? [Survey date: April 2011] [Answers: I do not know what QR technology is, 40.7%; Yes, 30.9%; No, 28.4%]

Mobile UPCommons[9]

One of SBDUPC's most important collections is UPCommons, a website containing the university's main institutional repositories and offering open access to much of UPC's scientific and academic output. The mobile version of the site offers access to all this material from mobile devices. It consists of a main website offering links to mobile versions of the main institutional repositories, as well as a search engine allowing users to search the material in those repositories. Access is currently available to the mobile versions of four UPC repositories:[10]

- E-prints UPC [UPC E-prints]: Repository consisting of the scientific documents generated by the university's faculty members and research community as a result of their research.
- Treballsacadèmics UPC [UPC Academic papers]: Comprises a largeshare of the final degree papers and master's theses written by UPC students.

- Revistesicongressos UPC [UPC Journals and conferences]: Open access journals published by the university and materials from conferences held there.
- Videoteca UPC [UPC Video library]: Repository of academic and institutional videos made at the university.

Each of these mobile websites has three sections:

- Novetats [What's new?]: The latest additions to the website (users have the option of subscribing to the website's RSS feed).
- Comunitats [Communities]: List of the communities and collections to which the materials belong.
- Cercador [Search engine]: Tool for searching the repository in question. Searches can be performed by title, author, subject, date, abstract and URI (URL of the file with the material's metadata).

These repositories, like the original ones, offer open access to full-text versions of all content under creative commons licences.

Like the other mobile websites, this one was developed by the library itself. However, it takes its data and materials from the UPCommons website, which is based on DSpace software. These data are imported daily to the mobile version through techniques such as screen scraping (to import communities and collections) or the OAI-PMH protocol (to import the material's metadata), which all these UPC repositories have.

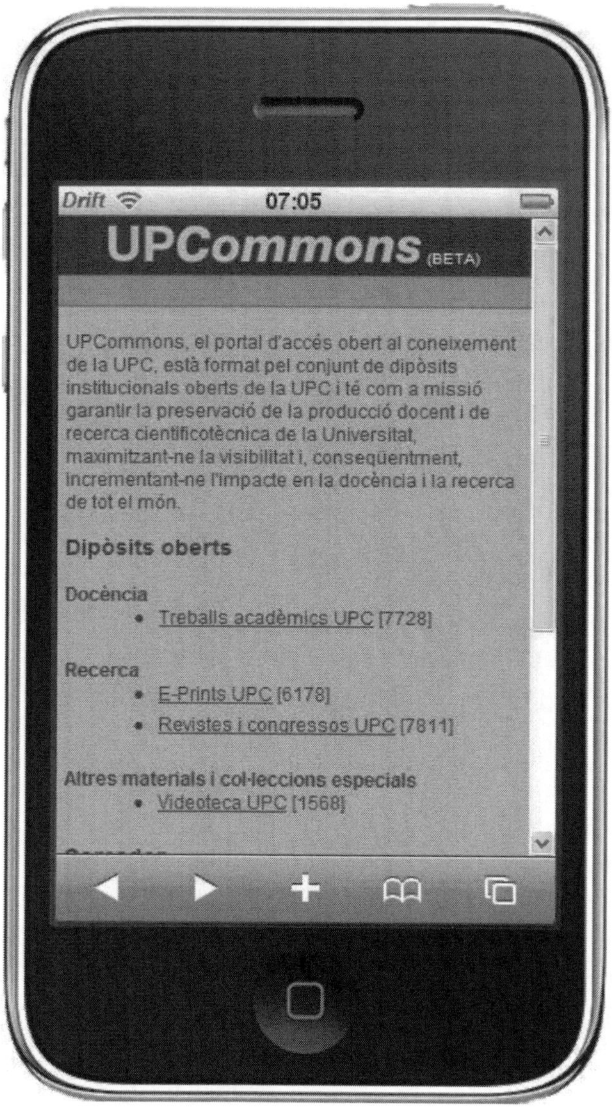

Figure 2. UPCommons mobile site.

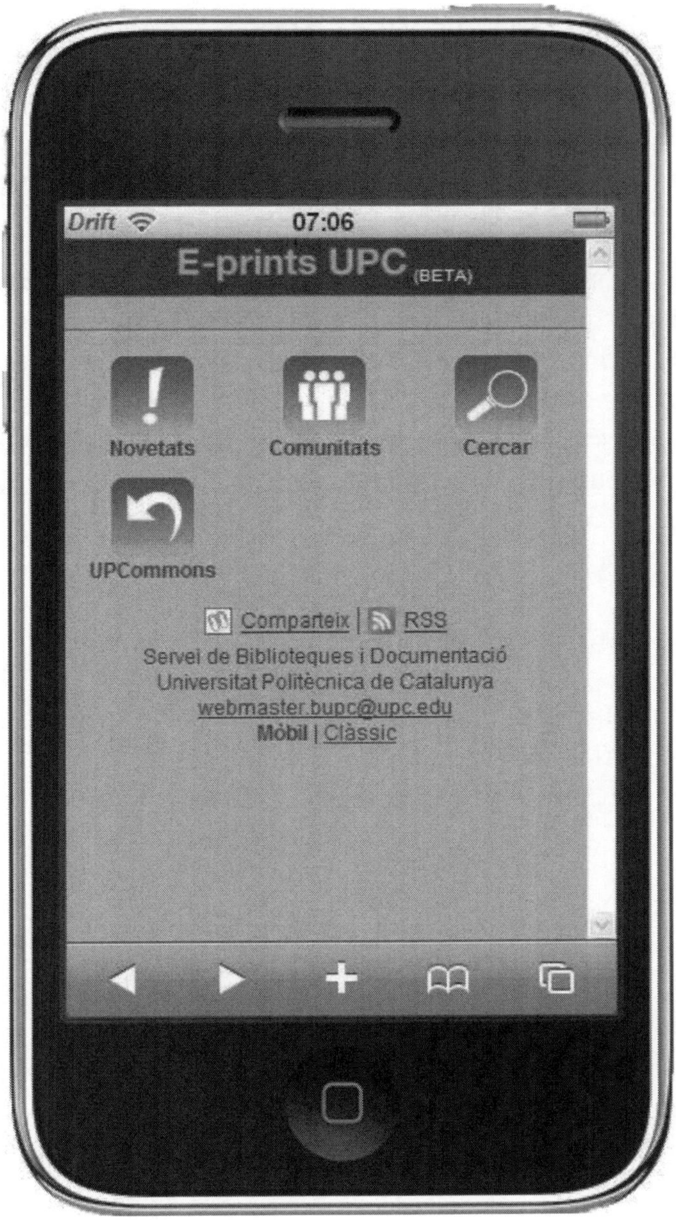

Figure 3. E-Prints UPC mobile site.

UPC's Mobile Catalogue (OPAC)[11]

The objective of the mobile online public-access catalogue (OPAC) developed by BRGF and SBDUPC is to allow users to search the UPC library system's catalogue from their mobile device under the same conditions as if they were using a computer.

The UPC library system's catalogue uses Innovative Interface's Millennium software (version 2009B1.2). This software optionally includes a module that makes it possible to create a mobile version of the catalogue. However, this module must be purchased separately from the subscription version and, after studying the costs, we decided to develop our own software for the mobile version of the UPC's catalogue.

The full application consists of two modules:

- Data mining: The full content of the Millennium catalogue is mined daily in XML format and, through internally developed software, uploaded into a MySQL database.
- The website interface: Developed using the same programming language and technologies as were used for the mobile website, this module is basically run against the aforementioned MySQL database. It performs real-time Millennium OPAC queries to determine the status of each copy in the bibliographic records. This function is quite important, because the daily data mining does not provide this information.

Together these two elements — the data mining module and the website interface — allowed us to create our own mobile version of the UPC libraries' OPAC, comprised of three main sections:

- Cerca[Search], available in two modes: a simple search, consisting of a text field in which users can enter their search terms, and an advanced search, in which they can also specify the search field (title, author, etc.) and limit the search to a specific library. The results show the main fields of each record (title, author and publisher), as well as the available copies at each UPC library. Optionally, a specific icon can be used to access the full record of each document and to see all copies. This is useful for determining the exact location of each copy in a given record.

- Biblios[Libraries]: This section allows users to view basic information about any UPC library, including its address, opening hours, calendar, a map and the telephone number.

- Ajuda[Help]: Brief instruction manual on how to use the mobile catalogue and documentation regarding the meaning of the different icons and colours used on the website.

Thus, as with the query function in the traditional OPAC, the mobile catalogue can be used to access the UPC libraries' subscription e-resources: in addition to individually catalogued e-books, the mobile OPAC enables access to e-journals, databases and reference e-books. Providers and publishers are also making a considerable effort to adapt their content to mobile devices and, at present, some of the key resources at UPC enabling convenient access from mobile devices include: IEEExplore, AIP/APS (the American Institute of Physics and the American Physical Society), ACS (American Chemical Society), Factiva, Emerald, Ebscohost, O'Reilly e-books, the journals Nature and Science, and the citation manager Refworks. Recently, UPC has also been working to offer content published with Edicions UPC in electronic format through the Iniciativa Digital Politècnica [UPC Digital Initiative] project.[12]

Subscription resources at UPC requiring authentication to be viewed can be accessed from mobile devices via a proxy server. Consequently, these resources can be accessed by any WiFi connection, including those that do not belong to the university itself.

Figure 4. UPC's mobile OPAC.

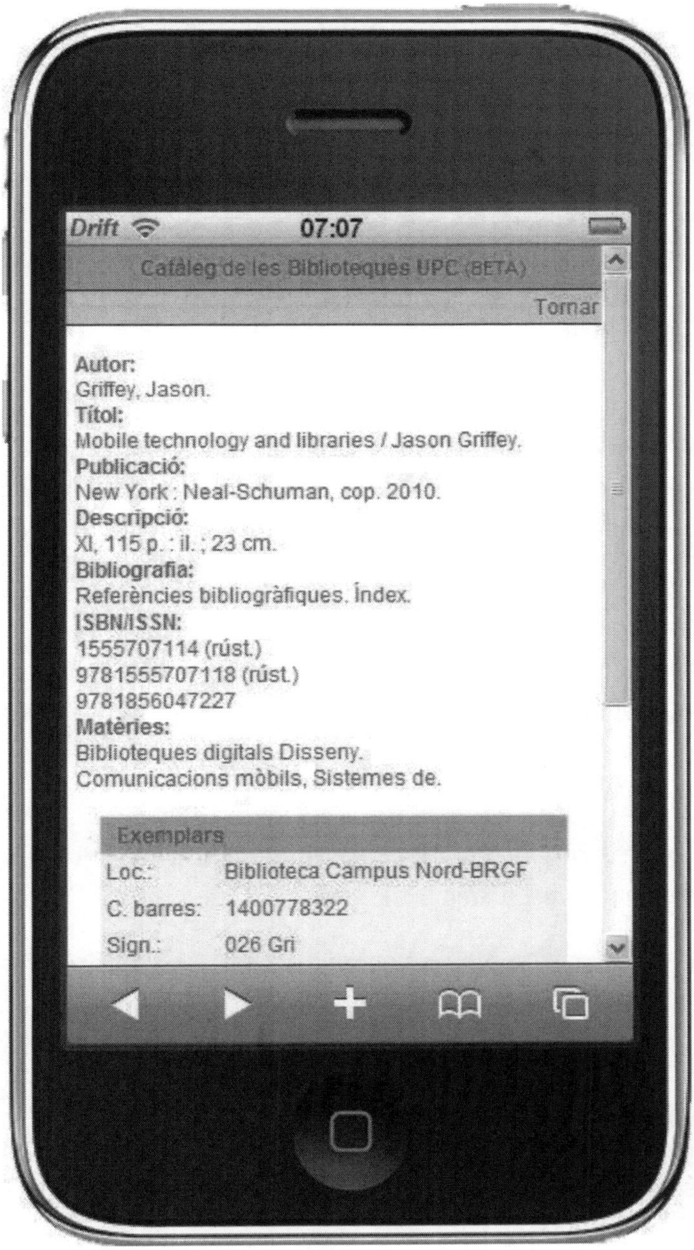

Figure 5. Bibliographic record on the mobile OPAC.

Mobile version of u-win **(video games at BRGF)** [13]

u-win is a service intended to broaden the library's traditional portfolio by including gaming. BRGF believes in the educational power of video games and considers them to be tools with great potential, for both faculty and students. It moreover believes that these tools can pave the way to a more social use of the library, in which there is room for recreation in addition to the university campus's traditional academic activities.

UPC offers postgraduate programmes and individual courses on video game development, in which students develop games as part of their coursework. u-win is used both to preserve and disseminate a selection of these games and to offer access to video games that were not developed at UPC, but which BRGF recommends.

Notwithstanding the above, besides a virtual platform, u-win is a physical space, located on the ground floor of BRGF and properly equipped. There, patrons can play some of today's most popular games on a PC or Nintendo Wii in a laid-backsetting fitted out with comfortable furniture.

u-win's offer is rounded out by two Nintendo DSi consoles, which can be checked out by users under the same conditions as the laptops.

BRGF has developed a mobile version of the u-win website offering the same features as the traditional version. However, the mobile version only includes games that have been adapted to mobile devices — in order not to frustrate its users.

u-win mòbil, the mobile version of the website, has the following sections:

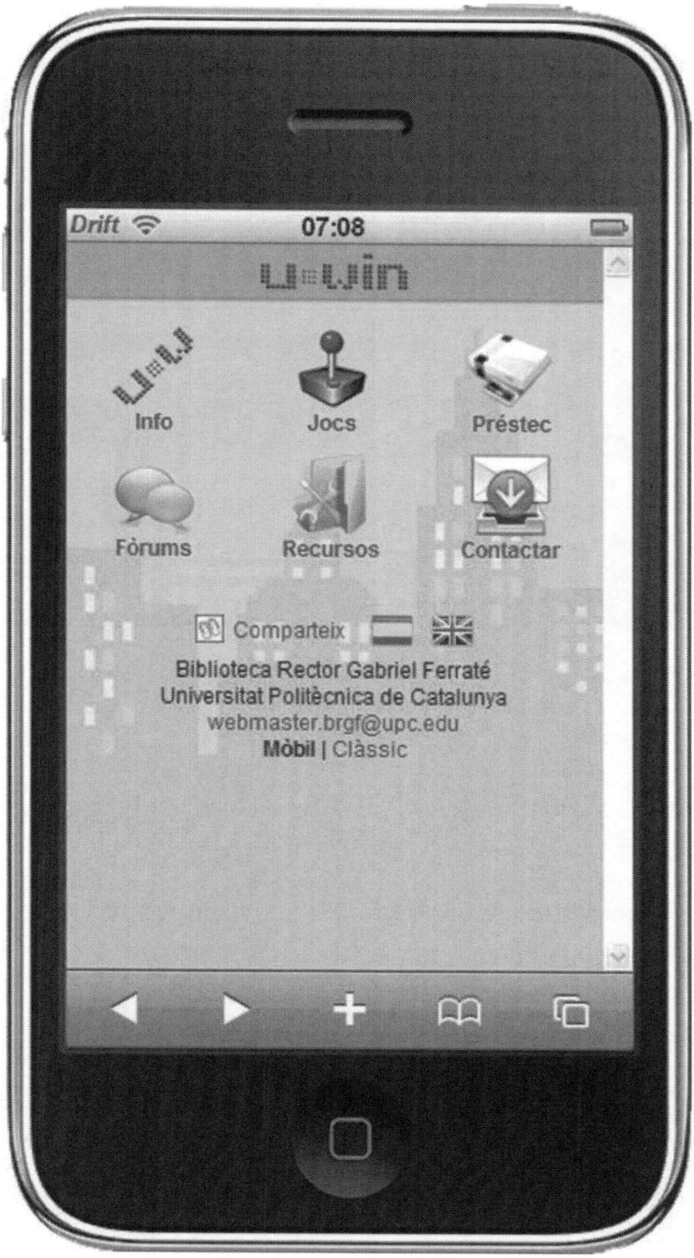

Figure 6. Mobile version of the u-win website.

Figure 7. Physical u-win space.

- Info [About]: Brief description of the website and the onsite service offered at BRGF.
- Jocs [Games]: Access to a search engine to find games with a mobile version. Searches can be performed by title, type (free, UPC or commercial) and category (action, sport, strategy, skill, combat, platforms, puzzles and shooter simulator).
- Préstec [Loan service]: Information on the console loan service, which includes two Nintendo DSi consoles and one room for playing a Nintendo Wii, in addition to the games themselves.
- Fòrums [Forums]: Area for registered users to discuss video games, programming, etc.

- Recursos [Resources]: Link to a selection of recommended reading, software and other online resources related to video game design and programming.
- Contactar [Contact us]: This option allows users to send proposals for new games and resources to the library and submit any queries they may have to the library staff.

The Use of QR Codes to Deliver Information to MobileDevices[14]

A Quick Response, or QR, code is a 2D image that can be used to store much more textual information than a traditional barcode. Initially, QR codes were created to manage factory stocks. However, over time, their use has become more and more oriented towards mobile phone users. The main advantage of a QR code is that, when a mobile phone with the necessary reader software installed is used to take a picture of it, the data it stores are automatically captured and transferred to the phone. There is thus no need to enter textual data manually using a mobile device's keyboard or to write it down on paper.

Currently, QR codes are used at BRGF in the following ways:

- On posters and signs to promote new services: The code contains the URL of the new website or service, primarily so that users do not have to remember it and can add it to their favourites or recommend it to others, but also occasionally to offer supplementary information about the website or service being announced.
- For user registration in the ClickArea: The code contains the username and password needed to access the library's computer classroom system. Thus, users do not need to keep track of any document with this information.

- To process suggestions for improving services sent by users to the library's institutional e-mail address using an adapted form.
- To confirm the reservation of a group study room: The code contains the date and time of the reservation of the group study room made by the user. It thus helps users to remember the details of their reservations.

Because they are easy to implement, and given that users are increasingly familiar with them, as will be seen in the section on possible future mobile services at BRGF, the use of QR codes is expected to register a clearly upward trend in the library's near future.

Organisation, Funding and Technological Image of the Library

The UPC library system's organisation chart has an unusual structure for a library service and is based on three operating units intended to meet the information and service needs of different user groups: the Learning Services Unit, the Research Services Unit and the Digital Services Unit.

Needless to say, technological development falls within the purview of the Digital Services Unit. In recent years, BRGF has tried to strengthen this unit in numerous ways:

- Providing it with a manager exclusively responsible for digital activities.
- Engaging a full-time computer engineer.
- Collaborating closely with the ICT staff of SBDUPC's General Services and with the staff of other libraries, as appropriate.

Figure 8. Sign with a QR code.

- Exercising the option of taking on interns from the two schools on campus dedicated to ICT: FIB and ETSETB.

At the same time, a permanent task force has been set up to monitor and propose new technology applications, especially ones with a mobile component. This task force operates in parallel to the Digital Services Unit and is made up of people especially interested in such applications from all the library's units. Above all, the task force members, like most of the library's staff, subscribe to information sources on technology and libraries in order to stay abreast of the latest developments in the field.

From an organisational point of view, it was the measures taken to strengthen the Digital Services Unit, together with the involvement of people from other units and the interest in determining users' opinions and keeping them satisfied by responding to their suggestions and comments, that allowed the library to develop the portfolio of technological services that has come to define it.

As for investment, in addition to the time and effort invested by the staff, it should be noted that the budget structure and budget management techniques of academic libraries do not tend to offer much leeway when it comes to investing money in these types of services, especially in software, which tends to be proprietary and to be priced beyond libraries' means, or, at least, beyond the UPC libraries' means.

Therefore, SBDUPC and BRGF have often chosen to:

- work with open software.
- develop in-house products with their own IT staff.

This policy has allowed us to create a range of products and services, the most significant example being the set of institutional repositories encompassed by UPCommons. While some of these products may not have as many features as others available on the market, they always allow us to set our own priorities in terms of what they need to offer users and, above all, they allow us to make our ideas a reality. To this end, all features of BRGF's mobile services have been developed by members of the library and SBDUPC staffs, who have adapted features that had already been developed for non-mobile purposes or programmed new ones in order to allow us to provide a mobile service. We have thus been able to offer our users a suite of mobile services without spending a single euro beyond our regular staff costs.

Separately, it must also be borne in mind that making the library portfolio available on mobile devices, and, in general, wherever users may be, lends the libraries an image of ubiquity, service orientation and modernity that is invaluable at a time when libraries are so at risk of being stripped of their traditional role as intermediaries in the information environment. Users, especially students, must not be left with the notion of libraries as mere storage places for books, or libraries will definitively disappear from the world of information and digital services. We must invest as many resources and as much imagination as possible in the commitment to technology and in the identification of libraries as part of this commitment.

However, in an environment that includes many expert technology users, it is difficult to determine whether technology is a means or an end in itself in the effort to get users' attention. BRGF strives to integrate the latest technology and available devices into what might be considered a more traditional library service. Thus, signs are enriched with QR codes, a space is fitted out to play video games, and a loan service is set up for e-readers, Nintendo DSs and iPads. There is always an underlying service or collection, or an educational or pedagogical rationale, to justify the use or acquisition of technology, but we also want the library to be seen as a media centre within the UPC's North Campus, a show case for technology and an engine for the application of new technologies to everyday life. Undoubtedly, this will enhance the library's reputation among users who arrive thinking that it is little more than a warehouse for books and is unlikely to have anything more to offer them. Perhaps users will be enticed to visit after hearing something about video games or tablets and will then fall into the dynamic of using BRGF's other resources and services, too.

Possible Future Developments at the UPC Libraries

Although the products developed to date already cover the main features we had hoped to offer our users, BRGF and SBDUPC remain firmly committed to adapting new services to mobile devices. To this end, new objectives and future lines of action have been established, first, with a view to enhancing current products:

- Enlarge the mobile version of the catalogue of UPC's institutional repositories by incorporating other repositories, such as the Open Course Ware repository or the catalogue of UPC doctoral theses included in the TDX repository managed by the Consortium of Catalan University Libraries (CBUC).
- Allow users to access their personal catalogue accounts (my catalogue) from the mobile catalogue so that they can take care of their personal business, view information on their loans, make renewals, etc.
- Help to expand the mobile website model to include the other libraries in the UPC system.
- Enable the mobile website to offer essentially customised services, including options such as allowing users to manage their group study room reservations or their login details for BRGF's ClickArea, among others.
- Improve the performance of all mobile websites, with special emphasis on search engines, and redesign certain interfaces to

make them more user-friendly in order to make them quicker and more convenient for users to browse.

- Allow videos from UPC's video library to be played in real time on devices such as the iPhone or Android, along the lines of the YouTube model, so that users do not have to download an entire video onto their device in order to watch it.

- Add covers and shelf maps (virtual shelves) to the mobile catalogue search results to make it easier for end users to find the documents they are looking for at UPC libraries.

- Adapt the library's themed websites — jazz, science fiction, Catalan poetry, and human and sustainable development — to the mobile version of the library's site.

- Increase the number of features that can be accessed through QR codes, enabling access to resources — identification of staff and functions at the BRGF, maps of the building's different spaces and facilities, etc. — or enriched access to collections — sound and image archives, journal abstracts, specific information about documents in the OPAC, links to external websites, etc.

- Second, efforts will be made to continue to work to develop new services and applications in this sphere. Some future possibilities include:

- Expand the use of text message notifications, e.g. to include reminders of users' room reservations, information on access to the library's services or to let users know when a document they have requested through the Document Retrieval Service has arrived.

- Implement a system of Bluetooth marketing campaigns aimed at further promoting the UPC libraries' new mobile features and services.

- Work on augmented reality projects that make it possible to find and enrich information on the spaces, equipment and documents available at the BRGF building.
- Implement a chat function to allow users to submit real-time queries to the BRGF's librarians.
- Given the projected dramatic growth of tablets among UPC users, explore the possibility of adapting mobile interfaces to tablets, too.
- Study the possibility and suitability of offering our own apps.
- In short, BRGF plans to continue to engage in intense work in the sphere of mobile devices, conceiving of and creating new products and services and evaluating their future evolution so as to bring the UPC library system's content and services closer to its users by means of their mobile devices.

CONCLUSIONS

Academic libraries have historically offered their services onsite, even when their portfolio was essentially technological. Equipment, facilities and the transformation thereof continue to be key to academic libraries' success. However, with the rise of the Internet, remote access to services has been important for years, and today we are immersed in a new context, in which communication, service and commercial interactions are increasingly conducted with mobile devices — indeed, some of these interactions can only be conducted with such devices. In this environment, it seems not only appropriate but indeed critical to develop new channels aimed at providing mobile services. In the words of M.L. Jacobs (2009), 'along with a new generation of technology, we have a new generation of users (...) [the] ING, the Information Now Generation. How better to describe a group of users that expects every thing at their fingertips regardless of where they are? And it is time for us to make and open mobile gates; there is no meeting someone half way to remain relevant'.

The Rector Gabriel Ferraté Library is located in a highly technological environment and considers its technological services to be its hallmark. Within this portfolio, BRGF has made mobile developments a clear priority in recent times and, through an organisation adapted to the strategic challenges facing the UPC Library and Documentation Service, has managed to develop a portfolio of mobile services that increase the means of accessing the library's equipment and information resources. Sometimes, this has been done even knowing that the time was not yet ripe for a given service or device, that the technology or concept was still too new. For example, the decision to create BRGF's mobile website was made at perhaps not the most opportune time, as Internet-enabled mobile devices had not yet taken root. Nevertheless, in light of current market trends, such devices — and, more specifically, smartphones with wide screens — were expected to become firmly entrenched and a part of users' daily lives and, consequently, the decision was made to invest in the mobile website project despite the potential risks. Moreover, as noted above, the projected development cost was not too high, and the consensus was that, despite the risk, it was worth investing in implementing the new version of the website, which, in any case, could be considered a pilot test for assessing the acceptance and success of such products among the library's users. The passage of time confirmed these predictions and today, nearly all users have an Internet-enabled mobile phone and the use of mobile services and applications is increasingly common, especially the use of those services shared by all UPC libraries. Consequently, the library has continued to develop new mobile projects, always for the purpose of bringing the library closer to users and facilitating and making more convenient their interaction with it. Furthermore, we plan to improve and expand these developments in the near future with a series of new features — explained above — that will simultaneously reinforce the idea of the library as a pioneer in the use of new technologies.

The UPC libraries consider being ahead of the curve to be a good strategy in general. At present, the commitment to mobile services is valued not so much in terms of the immediate use made of them — which, in any event, seems likely — as in terms of the positioning that such services imply. BRGF is convinced of the usefulness of this modus operandi and is keen to explore new possibilities to improve and expand the mobile offer in terms of access to content, additional and more customisable services, and its adaptation to the new devices that are constantly appearing on the market.

When it comes to mobile services, learning by doing is the UPC library system's motto.

ACKNOWLEDGEMENTS

The authors wish to thank Ms Margarita CeñaCosials of the Andorra National Archive, for her collaboration and the information and ideas she provided throughout the writing of this paper.

REFERENCES

1. American Library Association (2008): Games and gaming. Available at: http://connect.ala.org/gaming [Accessed: 04/05/2011].

2. Arroyo-Vázquez, N. (2009): 'Web móvil y bibliotecas', El Profesional de la información, 18(2), pp. 129–136.

3. Castells, M. ... [et al.] (2007): Comunicación móvil y sociedad: una perspectiva global. 2ª ed. Barcelona: Ariel, Fundación Telefónica. 475 p. ISBN: 9788434453272.

4. Codina Vila, M., A. Pérez Galvez and J. Clavero Campos (2010): 'Mobile services in the Rector Gabriel Ferraté Library, Technical University of Catalonia', Reference Services Review, 38(2), pp. 321–334.

5. Comisión del Mercado de las Telecomunicaciones (2010): Nota Mensual: Enero 2010. Available at: http://www.cmt.es/en/ publicaciones/anexos/ 100308_ NM_ enero010. pdf [Accessed: 04/05/2011].

6. Davidson, M. (2005): Make your site mobile-friendly in two minutes. Available at:www.mikeindustries.com/ blog/archive/2005/07/make-your-site-mobile-friendly[Accessed: 04/05/2011].

7. Federación de Gremios de Editores de España (2010): Informe Barómetro de Hábitos de Lectura y Compra de Libros en 2010. Available at: http://www. federacioneditores.org/0_Resources/Documentos/NP_Lectura2010_V3. pdf[Accessed: 04/05/2011].

8. Fundación Teléfonica (2010): La Sociedad de la información en España 2010. [S.l.]: Ariel, Fundación Telefónica. Available at: http://e-libros.fundacion .telefonica.com/sie10/ [Accessed: 04/05/2011].

9. Hanson, C.W. (2011): Libraries and Mobile Services. Chicago, IL: American Library Association. 34 p. Library Technology Reports, 47(2). ISBN 978-0-8389-5830-8.

10. Hassan, Y.; F.J. Martín Fernández and G. Iazza (2004): 'Diseño Web centrado en el usuario: usabilidad y arquitectura de la información'. Hipertext. net. No. 2. Available at: http://www.hipertext.net/ web/pag206.htm [Accessed: 04/05/2011].

11. Huwe, T.K. (2009): 'Thin client, meet the mobile future', Computers in Libraries, 29(5), pp. 22–24.

12. Instituto Nacional de Estadística (2010): Cifras de población referidas al 01/01/2010 Real Decreto 1612/2010, de 7 de diciembre. Available at: http://www.ine.es/jaxi/menu.do?type=pcaxis&path=%2Ft20%2Fe260%2Fa2010%2F&file=pcaxis&n=&L=0 [Accessed: 04/05/2011].

13. Jacobs, M.L. (2009): 'Libraries and the mobile revolution: remediation=relevance',Reference Services Review, 37(7), pp. 286–290.

14. Johnson, S. (2005): Everything bad is good for you: how popular culture is making us smarter. New York: Penguin Books. 238 p. ISBN: 9780141018683.

15. Kolb, L. (2008): Toys to Tools: Connecting Student Cell Phones to Education. Eugene, OR: International Society for Technology in Education. 230 p. ISBN 9781564842473.

16. Kroski, E. (2008): 'On the move with the mobile web: libraries and mobile technologies'. Chicago, IL: American Library Association. 48 p. Library Technology Reports, 44(5). ISBN 978-0-8389-5795-0. Available at: http://eprints.rclis.org/bitstream/10760/12463/1/mobile_web_ltr.pdf/ [Accessed: 04/05/2011].

17. Liston, S. (2009): 'OPACs and the mobile revolution', Computers in Libraries, 29(5), pp. 6–8, 10–11, 42–47.

18. McGonigal, J. (2011): Reality is broken: why games make us better and how they can change the world. New York: Penguin. 388 p. ISBN: 9781594202858.

19. McKiernan, G. (2009): Mobile libraries. Available at: http://mobile-libraries.blogspot.com/ [Accessed: 04/05/ 2011].

20. Midwestern Mac, LLC (2009): Adapting your web site for the iPhone. Available at:www.midwesternmac. com/tutorial/2009-01-13/adapting-your-Website-iphone[Accessed: 04/05/2011].

21. Ofcom (2009): Digital Lifestyles: Young adults aged 16-24. Available at: http://stakeholders.ofcom.org.uk/ binaries/research/media-literacy/young_digital_lifestyles .pdf [Accessed: 04/05/2011].

22. Ofcom (2011): The Communications Market 2010 (August). Available at: http://stakeholders.ofcom. org.uk/market-data-research/market-data/ communications- market-reports/cmr10/uk/ [Accessed: 04/05/2011].

23. Ruadhan (2008): Lightweight device-detection in ASP. Available at: http://mobiforge.com/developing /story/lightweight-device-detection-asp [Accessed: 04/05/2011].

24. Seeholzer, J. and J.A. Salem (2010): 'Library on the Go: A Focus Group Study of the Mobile Web and the Academic Library', College and Research Libraries, 27(19–20), pp. 9–20.

25. Serrano-Muñoz, J. (2000): Acceso a la Biblioteca de la UOC por medio de la telefonía móvil. Available at:http://biblioteca. uoc.edu/cgi-bin/pass/byteserver.pl /docs _elec/ponencies/3538.pdf [Accessed: 04/05/ 2011].

26. The Nielsen Company (2010): Mobile Youth Around the World: December 2010.Availableat: http://www.nielsen.

com/content/dam/ corporate/us/en/ reportsownloads /2010%20Reports/Nielsen-Mobile-Youth-Around-The-World-Dec-2010.pdf [Accessed: 04/05/2011].

2

CHAPTER

The Future of the Map Library and Map Librarian

John N. Moore

Collections Officer/Subject Librarian, Physical Sciences Faculties Support Team, Glasgow University Library, Hillhead Street, Glasgow G12 8QE, UK

ABSTRACT

Interest in maps and cartography is on the rise, partly spurred by new technologies such as Google Earth and Google Maps. But the new users, welcome as they are, do bring new challenges in terms of ease of use and speed of delivery. Therefore, LIBER's Groupe des Cartothécaires devoted its biennial conference to state-of-the-art technologies and concepts such as new retrieval techniques, Web 2.0 content, links with geospatial databases and FRBRisation to investigate their potential for and impact on map libraries and map librarians — whilst concluding that the particular expertise of a map librarian is as indispensable as ever. John Moore, the Group's Secretary, reports on the conference.

Keywords: Groupe des Cartothécaires, map curatorship, map libraries/archives

In what turned out to be an innovative and stimulating programme, the Dutch organising committee of the 16th Conference of the Groupe des Cartothécaires had moved the timing of the congress to the beginning of the summer. Although this denied most delegates the opportunity of a break from their duties before attending the conference, the level of participation, discussion and interaction did not indicate that this was a problem. In fact, this was one of the most successful conferences of recent years and the organisers impressed many by the impact created by certain changes made to the usual pattern, the choice of papers presented and the overall arrangements for the week.

Our hosts in the Netherlands were the Amsterdam City Archives (SAA), supported by the libraries of the University of Amsterdam (UvA), the Vrije Universiteit (VU University), Wageningen University and Research Centre (WUR) and the Royal Tropical Institute (KIT). Members of the organising committee represented the host institutions, thereby ensuring a wide spread of interest and relevance.

As in past conferences, the high number of delegates (over sixty) from twenty European countries reflected a variety of backgrounds in libraries (national, university, public), archives (national, regional), national survey bodies and research institutions. On this occasion, a welcome return was made by representatives from Iceland, Ireland, Italy and Portugal. While the programme contained the usual mix of presentations on developments within particular areas of specialisation, the organising committee had put much thought into structuring a programme which introduced notable changes from past conferences. In particular, the various national reports on cartographic activity and organisation within thirteen of the countries represented were reorganised into part of a broader poster session in the Special Collections Department of Amsterdam University Library, thereby allowing individuals the opportunity to discuss presentations with presenters.

On this occasion, the conference theme was: 'The future of the map library and the map librarian', reflecting the concerns of many in the profession of the impact which significant technological change in methods of cartographic production and delivery is having on access, collections and expertise.

Prior to the beginning of the conference proper, some delegates enjoyed a very interesting historical cartographical tour of the centre of Amsterdam under the expert tutelage of Jan Werner from Amsterdam University Library and Secretary to the organising committee. Jan's infectious enthusiasm for the strong tradition of map-making within the city and his intimate knowledge of the close social and geographical inter-relationships between many of the key figures in the golden age of Dutch cartography reminded those fortunate enough to accompany him of the rich history of map-making in the Netherlands.

The conference began in the recently refurbished Film Theatre of Amsterdam City Archives and delegates were welcomed formally by Marc Hameleers, President of the organising committee to his home institution and the city. Marc stressed that this was the second visit by the Group to the Netherlands, having previously enjoyed a conference in The Hague in 1984. He also highlighted some of the changes introduced to the present congress, including the creation of discussion groups. Following on from this, a thought-provoking and relevant keynote address to the conference entitled 'Mapping out map libraries' was delivered by Ferjan Ormeling (Utrecht University). His paper considered maps as models of reality and, as his first illustration was the famous map illustrating Robert Louis Stevenson's 'Treasure Island', he had the interest of the audience from the outset. He reminded us that curators have the experience and knowledge of the reality within our collections to be able to utilise them and provide the metadata necessary for proper and wider disclosure. Such expertise is even more important in an age when the availability of cartographic material is affected by the price of files and the complexity of geospatial information packages. The future will require us to help users bridge the information gap, developing into a more consultative and educational role as interest in geographical data continues to grow. Curators are uniquely placed to blend the skills of treasure hunting and GIS knowledge, combined with an awareness of both data content and reader requirements into a package which is highly relevant to many in the map using community.

Following on from this vision, the conference turned to more practical and local experiences with presentations from three different backgrounds. Peter Korsgaard (National Survey and Cadastre, Denmark) reflected on what has happened in his own institution over the last two years where all new maps are digital and the archive is being digitised for Internet availability. Nonetheless, the knowledge of the map curator is essential in providing the relevant metadata for cataloguing systems and the user with information on the most appropriate maps, regardless of source. Chris Fleet then offered the audience a critical evaluation of some of the trends in new geographic retrieval technologies based on his own experience at the National Library of Scotland in retrieving historical online mapping. As ever, Chris's paper provided a reasoned and well considered reflection on tools for access and our unique position in being able to utilise these for the benefit of our users. There is no one answer to our users' variant needs and different situations will require different types of graphic. Nonetheless, certain trends are becoming more defined and with a prescient allusion to the papers delivered on the Friday of the conference, Chris referred to dynamic zoomable maps as an essential tool to the basic core of order and arrangement, and yet most users seem to prefer textual lists, emphasising the importance of a mix of information. Whatever develops in the near to distant future, map curators remain the best placed experts to fill the niche between the web as the sole information channel for libraries and the technological focus on the user. Embracing such as priorities will be a key future skill to provide a broader role. To conclude the morning session, Arnold

Bregt (Wageningen University and Research Centre) based his paper on the user's point of view, looking at the role maps and map collections play in e-government. The widening of the 'spatially aware' readership and a more demanding clientele, combined with the un-stated expectation of staff becoming guardians of the future, will result in a supportive and interpretative role for map room staff. Arnold stressed a constant theme of the conference that we need to become part of the spatial data infrastructure. A challenging but brighter future than many may have viewed before attending the conference.

Having a chance to reflect on this future over lunch, delegates were then challenged by Greetje de Jong, a professional facilitator, to participate in four discussion groups considering variant alternative futures. We were encouraged to look at future developments and what driving forces lay behind change, resulting in a breakdown of possible future scenarios based on the impact of technology and centralisation and our management of these forces. This was a new approach to encouraging delegates to contribute to the general discussion and, in certain groups, it stimulated much more debate than usual! Having cudgelled our brains into some better vision of what lies ahead, we then dispersed for dinner and a more reflective consideration of what we seek to do.

The second working session, Modern map-use in education, took place on Wednesday morning in the Vrije Universiteit Library. This followed an introductory visit to the Zuidas Information Centre which put in perspective the development of this new urban centre in south Amsterdam — an urban planner or sociologist's vision. Three papers formed the content of the morning session, chaired by the Group Secretary in (almost) full national costume — Martijn Storms (Leiden University Library) on 'Teaching a map collection', Nick Millea (Bodleian Library, Oxford University) on 'The Gough Map, the map library and the research community' and Wybren Verstegen (Faculty of Arts, Vrije Universiteit) on 'The Nature Scenery Act in the Netherlands and the reconstruction of past landed estates'. Once again, the papers seemed to dovetail together very well, despite a last minute change in the programme. As ever, Nick presented a polished exposition of pivotal work being done by the library on encouraging new research into the earliest surviving modern map of Britain. His illustration of the potential role for map librarians in identifying material for future research is one illustration of how institutions can realise the full potential of collections. Martijn gave the conference an overview of a master course he had run in the Book and Digital Media Studies programme at Leiden based on the Bodel Nijenhuis collection. He reminded us that Leiden does not have a department of geography and of the difficulties faced by introducing such material to students not necessarily as conversant with the concept of space as many of our more traditional users. Looking at collections from a researcher's point of view, Wybren provided the background and provenance to a collection of maps relating to a significant piece of environmental legislation in the Netherlands.

Delegates were then offered a choice of visits to the University Map Collection, the MediaXperience and demonstration map table and to the roof of the building where they could enjoy a view to the north over the city, reminding several of us of the Paris conference and the view from the Tolbiac building. Once again, our hosts treated us to lunch before braving a heavy downpour of rain on the return to the Special Collections of Amsterdam University Library for the afternoon session. This gave delegates the opportunity to view the national reports and other poster sessions, which included strong representation from both Poland and Russia on the provision of cartographic information in both countries, as well as updates from Wolfgang Crom (Berlin State Library) on Bibliographica Cartographica Online and Annick Anceau (University of Liège) on her work first presented in Paris two years ago. This continuity is a most valuable development in the programme arrangement for it allows delegates to keep abreast of the continuity of major projects and provides a pattern of information input which can be used in the individual situation. In addition to the posters, delegates could view the major exhibition of Blaeu's 'Atlas Maior', again underlining the immense wealth of historical Dutch cartography. Several expressed their envy of the generous exhibition space in the newly refurbished library. As well as the exhibitions, the party were offered tours of the Special Collections and map room, where there were demonstrations of new digitising and printing equipment, access to the holdings of the Royal Tropical Institute and Delft Technical University's Geodesk.

Those who still had an appetite to view paper maps and atlases to be fed were well nourished by a market display by rare book sellers, conservators and cartographic interest groups. While few curators are likely to be in the fortunate position of having funds to purchase many of the treasures on display, it is a timely reminder of two key elements in this area of interest — the invaluable bibliographic knowledge many dealers have of maps and atlases and the high market value of many of the treasures under our care. The day was well rounded by an enjoyable and relaxing canal cruise generously provided by the Municipality of Amsterdam.

'New methods for the version of geographic and cartographic data' was the title of the third working session held on the Thursday morning in the Forum Library of Wageningen University and Research Centre. This was ably chaired by Nick Millea (Bodleian Library, Oxford) and consisted of three papers by Marco van Egmond (Utrecht University Library), Ruth Kalf (Université catholique de Louvain) and Anda Zalite (National Library of Latvia). With some reference to the work he first introduced to the congress in Paris, Marco considered the challenging issues of archiving and access to digital cartographic data, in particular the difficulties of data migration and the need for rigorous cataloguing as well as metadata provision for the digital document. Once more, feeding into a general trend of the conference, Marco suggested that there was a place for international cooperation in providing high quality records by participating institutions — yet another example of the changing role of the map librarian to one of information manager.

Ruth then introduced a possible cataloguing solution in her paper 'FRBR: an opportunity for map collections and map users?' Functional requirements for bibliographic records is an IFLA 'entity-relationship' model for a more complex but richer catalogue, grouping collections by work as an intellectual creation. While still in the development stage, this has the potential for a revolutionary approach to open up our collections to new possibilities and user groups. It is to be hoped that Ruth will have the opportunity to keep the Group updated on the progress of this new concept. Anda's paper, with the eye-catching title of 'The way to the castle of light' discussed the role of the Department of Cartographic Materials at the National Library in the dissemination of geodata. In her quiet but effective way, she detailed the rapid technical developments in Latvia and their impact on the work of the map librarian, particularly in the planning of the new National Library building. Here is an excellent example of how we must look to trends to identify what kind of future there might be for such specialised collections and the support/promotional role that staff can play should not be underestimated.

During the lunch interval, the Group Board met for a lively discussion on progressing many of the ideas beginning to crystallise into more definite trends and, as Secretary, it was heartening to see such enthusiasm being expressed. The afternoon session was handed over to our hosts at Wageningen who provided us with updates on the work taking place there. Frans Rip introduced his audience to GeoDesk and its symbiotic relationship with the map library. Both use metadata and provide services to their users suggesting a sharing of knowledge as one potential development of the thorny problem faced in institutions where experience of GIS software is limited to the few. By combing expertise from both areas, there are opportunities to open up collections to provide far wider access. Henk Kramer's presentation concentrated on historical maps and GIS databases with particular reference to land-use. His overview of the work in extracting such information from the available topographic series gave an illustration of the facilities now available for investigating a time-series of documents with examples relevant to both the Netherlands and Europe in general.

The remainder of the visit to this innovative institution was dedicated to group tours of the Library's Special Collections enthusiastically led by Liesbeth Missel and members of her staff. The tour included displays of the collections of aerial photography from the Second World War, a unique garden architecture archive and a maps exhibition entitled 'Wageningen: op de kaart, een typologie van kaarten'. In addition, delegates were offered a series of short demonstrations relating to the Alterra Map Database, ISRIC World soil library and maps, TopoXplorer and Map Table. It was rather like an Ideal Home exhibition for the would-be GIS/map librarian of tomorrow but again underlined the rapid and divergent technological changes impacting on the profession, in particular the facilities offered for better disclosure and wider access.

Somewhat sated by the variety of presentations which gave the congress much to consider, the delegates travelled to the Veluwe area for the conference dinner where the inner person was nourished in what was a converted barn. There was ample opportunity to discuss many of the issues raised before the journey back to Amsterdam. Re-vitalised by this, the fourth working session 'Visual access to map information' took place on Friday morning in the Koningszaal of the Amsterdam City Archives. Henrik Dupont (Royal Library, Copenhagen) chaired what turned out to be a grand finale to our week of deliberations with a series of four highly relevant papers discussing tools to access cartographic resources. Peter Levi (Royal Tropical Institute, Amsterdam and conference webmaster) introduced the work done to digitise the 12,000 maps from the cultural heritage collection covering the period 1850–1950. Peter provided some very practical guidance on organising such a project but stressed how such work opens up new possibilities for collections and map library staff. Following from this, and with echoes of ideas which had drifted around at the previous conference, Wolfgang Crom (Staatsbibliothek zu Berlin) reported on his progress with a project to develop a web-based graphic cataloguing and search system for maps at the German Research Foundation (GOKaRT). Taking a cooperative approach as a means to combat problems faced by understaffed collections, the scheme has a value and implications well beyond the national boundary, particularly if sufficient funding can be found for free international use.

A French contribution to a similar method of providing access to collections was the subject of the presentation by Jean-Luc Arnaud (CNRS-MMSH, Aix-en-Provence). CartoMundi is being developed as a website for cartographic resources based on an open network of members. Jean-Luc emphasised one of the major stumbling blocks for many a researcher — abundant production but poor recording of multi-sheet series. The solution which we appear to heading towards suggests cooperative work on accessible graphic indexes, central to disclosing our disparate collections. In effect, the issue of such web available tools struck some attendees as a natural progression from such cooperative catalogues as WorldCat or COPAC but dealing more effectively with the difficulties arising from multiple sheets in sets. The final presentation of the conference was given by Jose Borbinha (Instituto Superior Tecnico, Lisbon) and introduced the work being done in the DIGMAP project within the European Commission's eContentplus programme which seeks to enhance visual access to historic collections. While several national libraries are involved, the work has a much wider relevance to the promotion of all research/heritage collections, based on a specialised service using the libraries' own metadata.

There was much to think about in these four presentations, particularly as they seemed to dovetail into and build on earlier schemes to open up our collections (e.g. Toporama). It is this author's opinion that the present convergence of so much thinking and work on these projects gives the Groupe des Cartothécaires an ideal opportunity to engage with the key players and contribute effectively to co-ordinating activity for the greater benefit of all. A debate on the issues involved and some clear thinking on a way forward would be welcomed by the many.

With so much to think about, the official conference closed with the Friday afternoon session as a series of reports back from the brainstorming groups created earlier in the week. The groups each presented their vision of a future which would be influenced by the driving forces of centralisation/decentralisation and the variants between paper and digital collections. Each group had catchwords for their scenario, namely 'Maps come to you', 'Big and powerful', 'Paper is alive' and 'Human face of maps'. In truth, some of the presenters were caught out by this requirement to provide a reasoned report and this author must record his indebtedness to his Swiss collaborator for saving the day. Whatever else this session did, it threw up two significant features — that there is a very strong thread of the theatrical amongst the British contingent who took on their roles with gusto and, more importantly, we are passionate about the importance of map curators to the full promotion and development of the collections under our care.

With the more cerebral part of the proceedings over, many delegates enjoyed a visit to Amsterdam's Westerkerk on the Friday evening before a post-conference tour to West Friesland and the Markermeer area on the Saturday. Our guide for the morning, Jan Beenakker (University of Amsterdam) provided us with a geographical commentary of the area. We eventually halted at Enkhuizen, once a major port but now thriving as a tourist honey trap on the west coast of the IJsselmeer. We were taken on a walking tour of the medieval town, along its fortifications, around the harbour, and never far from the Enkhuizen coat-of-arms — the three herrings. Our organisers (Jan Beenakker and Jan Werner) excelled themselves by leading us to a café, and instructing us to sit next to a cake of our choice — amply proportioned gateaux abounded, and so we were presented with an opportunity to escape the hot sun for some well-earned refreshments.

Back on the bus, we headed west to Hoorn, where we were welcomed at the Westfries Museum and given a presentation by Onno Zaman, director of Pictura Imaginis BV, on the Frederick de Wit wall map digital restoration project. Enkhuizen town council owns five 17th-century de Wit maps, comprising an unusually complete set covering the whole world. Subsequent damage after a fire rendered the maps no longer fit for display, so a digital restoration project was set up. The Pictura team has managed to remove cracks and bring back faded colours, much of the background information coming from a second, less complete set of the series held at Amsterdam University Library.

An online presentation followed, enabling us to compare the original imagery with Pictura's recent efforts, and the results were most impressive.

More refreshments in the walled garden behind the museum set us up for a tantalisingly short stroll through the town to the harbour, where our arrival coincided with that of the Nieuwe Liefde, and we were welcomed on board for lunch and a sailing trip into the Markermeer. In glorious conditions, we plied south towards Volendam, dining on deck, and generally enjoying the wind in our hair (if not the ship's sails).

The Nieuwe Liefde was too large to land on the former island of Marken, so we were able to moor at Volendam, walk around the harbour, and take a launch across the water to Marken itself, now connected to the mainland, although the causeway is a considerable distance from Volendam, hence the need to travel by boat. Once on the 'island' two walking parties, escorted by the Werners, witnessed this strange land, much of it below sea level, with houses clustered together in 'werfs', best described as raised mounds of earth.

The weather closed in, so our return to the Nieuwe Liefde was taken below deck on the launch, then involved a quick sprint around the harbour to reach the dry of our ship's bar. Buffet dinner and drinks were served (above deck for those who profess a maritime tradition, down below for the majority), but as Amsterdam neared, the skies cleared, and we made our way under a tilting bridge and through a lock, and back into the River IJ. Unfortunately, on our return to Sumatrakade, another sailing ship flying the skull and crossbones had eased into our berth, so our skipper docked alongside the Soeverein, and a gangplank was fixed to enable us to make safe passage to shore.

The most difficult moment of the whole conference had been reached — when goodbyes had to be said to people who had become friends. However, we hope it is more a case of 'au revoir' and, while two years may seem be very long, the organising team from Estonia will certainly find it passing very quickly. Those non-sailors in the party gradually made their way into the city, whilst the residents stayed put to toast an excellently organised and very successful conference.

An extensive website for the conference, with related abstracts, power point presentations, national progress reports, posters, programme and photographs has been maintained by Peter Levi.

Websites Referred to in the Text

ACKNOWLEDGEMENTS

The author is much indebted to Nick Millea (Bodleian Library) and Ruth Kalf (Université catholique de Louvain) for their contributions to this report and to all who attended the conference for making it such a success.

Websites Referred to in the Text

Conference website, http://liber.gdc.2008.googlepages.com /home
Frederick de Wit maps at the Westfries Museum,
http://www.wfm.nl/ index.php?option=com_ content&task =view&id=221&Itemid=99

3

CHAPTER

Designing Libraries for Research Collaboration in the Network World: An Exploratory Study

Sheila Corrall

University of Pittsburgh, US

ABSTRACT

Advances in technology and changes in higher education and research are forcing libraries to rethink the services they offer to researchers and how they are presented. Librarians are moving into "higher end support" and adopting new service models based on "deep collaboration" with academic partners. The internal arrangements that form the design of an organization can determine whether it achieves its mission under particular conditions. The present study explores how libraries in 24 leading UK research universities are organizing resources and services to support the research enterprise. Qualitative data were collected from institutional websites and other public domain sources and were analysed using matrix techniques. The results show a trend away from integrated library and computing service organizations; variety in institutional reporting lines, but predominantly in large professional service groups; consistency in internal library groupings, but variation in portfolio and job titles; expansion of specialist positions, with new functional roles complementing traditional subject liaisons; and dedicated spaces, working groups, and integrated websites promoting boundary-spanning activities. The findings confirm and extend prior work and are being used to design a large scale international survey.

Keywords: management structures, organization design, research support, university libraries

1. INTRODUCTION

Developments in digital media, network technologies, research workflows, scholarly communication and funding policies are challenging academic libraries to respond with a wider array of services and facilities for researchers (**Bourg, Coleman, & Erway, 2009**; **Housewright, Schonfeld, & Wulfson, 2013a**; **Vinopal & McCormick, 2013**). Subject/liaison librarians have traditionally provided research services as part of their academic support role, but libraries are increasingly identifying research support as a specific area requiring additional co-ordination and strategic development:

*"Open-access publishing, institutional repositories, the need to co-ordinate collection management and storage, the increasing availability of information technology (IT) tools to help researchers with their work and, not least, the Research Excellence Framework (REF) have all pointed to the need for a coherent library strategy and response" (**Bradbury & Weightman, 2010, p. 65**).*

University library support for research has emerged as a strategic issue and generated studies investigating service developments, new roles, skills requirements (**Auckland, 2012**; **Corrall, Kennan, & Afzal, 2013**; **Covert-Vail & Collard, 2012**; **Jaguszewski & Williams, 2013**). The present paper offers a different perspective on the topic by examining library engagement with research and researchers through the lens of organizational design.

1.1 Background

There is a perceived requirement for libraries to develop more specialized interventions to provide point-of-need research support and advice around the whole knowledge creation cycle: from ideas generation and project conception, through data acquisition, manipulation and interpretation, to the deposit of results, publication of findings, and assessment of impact (**Larsen & Riis, 2012**;**MacColl & Jubb, 2011**; **Vaughan et al., 2013**). Libraries are accordingly moving into areas such as funding opportunities and grant writing, ethics review, data curation and repository management, poster design and conference hosting (physical and virtual), journal and monograph publishing, bibliometric evaluation and impact assessment (**Adema & Schmidt, 2010**; **Carroll, 2011a**;**Crow et al., 2012**; **Delserone, Kelly, & Kempf, 2010**; **Drummond & Wartho, 2009**;**Furlough, 2010**; **Healy, 2010**; **Herther, 2009**; **Lyon, 2012**; **McBain, Culshaw, & Walkley Hall, 2013**; **Vaughan et al., 2013**), alongside their established roles in collection stewardship, resource procurement and information literacy. The development of support for the management of research data in line with national policy is a particular concern (**Carlson & Kneale, 2011**; **Gold, 2007**;**Lynch, 2008**; **Lyon, 2012**).

Commentators increasingly assert that libraries need to reposition from support service to professional/scholarly partner and to transform their liaison librarians into embedded experts recognized as academic associates (**Carlson & Kneale, 2011**; **Duranceau, 2008**; **Fonseca & Viator, 2009**; **Gold, 2007**; **Vandegrift & Varner, 2013**), so that they can add value through "higher end research support" (**Corrall et al., 2013, p. 638**). **Vaughan et al. (2013, p. 313)** argue that service presentation is an important factor here: "By basing the service model on the research lifecycle, the library is then poised to be a partner through the entire process, not just at the bookends of research". The proposed shift from the "service-and-support" approach to a partnership model is significant. **Posner (2013, p. 45)** argues that contemporary digital humanities projects "do not need*supporters* – they need *collaborators",* explaining that libraries need to provide both infrastructure (tools, servers, etc.) and "intellectual labor" (knowledgeable librarians). **Vandegrift and Varner (2013, pp. 69, 76)** similarly explain how partnering scholars in digital humanities should go beyond collection building to content creation, by "making 'stuff'" (e.g., web sites, digitized collections, new tools), describing the model as "deep collaboration".

Others have questioned whether libraries have sufficient expertise for such roles, particularly the technical understanding and domain knowledge needed for data management and curation in different disciplines (**Gold, 2007**; **Lynch, 2008**). However, feedback from biomedical and science librarians collaborating as partners in a large-scale cross-disciplinary multi- institutional research project reveals that personal attitudes and interpersonal skills are the key factors for successful participation in team science, with the willingness to learn and adapt more important than a deep understanding of the technology (**García-Milian et al., 2013**). Sector organizations including Research Libraries UK, the Joint Information Systems Committee, and the Association of Research Libraries (ARL), have sponsored projects examining the skillsets needed for new or extended subject liaison and specialist roles (**Auckland, 2012**; **Cox, Verbaan, & Sen, 2012**; **Jaguszewski & Williams, 2013**). Library and information schools are also responding with modules, courses, specializations, internships, and institutes to prepare new professionals and update the existing workforce (**Creamer, Morales, Kafel, Crespo, & Martin, 2012**; **Harris-Pierce & Liu, 2012**;**Lee & Tibbo, 2011**; **Stanton et al., 2011**; **Yakel, Conway, Hedstrom, & Wallace, 2011**).

The model of the librarian as a "scholar-practitioner" and active researcher is being promoted in several institutions (**Giesecke, 2010**; **McBain et al., 2013**;**Schrader, Shiri, & Williamson, 2012**), and being used to bring people with PhDs but no library degrees into research support roles in "alt/ac" (alternative to the academy) positions, particularly in relation to digital humanities developments (**Brunner, 2010**; **Giesecke, 2010**; **Nowviskie, 2013**; **Shore, 2012**). Even so, **MacColl and Jubb (2011, p. 10)** point out that despite recognizing their need for support, "researchers have little interest in the support services libraries have built for them". **Jahnke and Asher (2012, p. 16)** similarly report that "Few researchers see the library as a partner". Two large-scale surveys of academic faculty conducted in the US and UK confirm this message, with only half of the US respondents and one-third of the UK sample rating the library role in supporting their research productivity as "very important", placing this role as the least important of the six service areas specified (**Housewright et al., 2013a**; **Housewright, Schonfeld & Wulfson, 2013b**).

Some research libraries are responding to the challenge by moving beyond alignment with academic departments or schools to focus provision of in-depth information advice at the more specialized and dynamic level of research groups and projects, particularly in the biomedical and health sciences area (**Carlson & Kneale, 2011**; **Federer, 2013**; **Greyson & Surette, 2013**). Another significant trend is the creation of specialist positions dedicated to research support, either exclusively or in addition to a subject librarian role (**Bradbury & Weightman, 2010**; **Young & Lund, 2008**), including positions coordinating support for graduate students (**Covert-Vail & Collard, 2012**). The 2013 Ithaka survey of US library directors indicates that almost two-thirds of respondents in doctoral institutions expect to increase staffing for specialized faculty research support (**Long & Schonfeld, 2014**). Others have decided that traditional structures, such as the subject-specialist information consultant model, are no longer viable, and have reorganized their staffing to align with the broad institutional missions of research and teaching, or adopted business enterprise structures, with teams dedicated to service development and innovation (**Andrade & Zaghloul, 2010**;**Cotta-Schønberg, 2007**; **Covert-Vail & Collard, 2012**; **Drummond & Wartho, 2009**; **Franklin, 2009**; **Manchester University Library, 2012**; **Nutefall & Chadwell, 2012**).

1.2 Related work

Organization design has not received due attention in the debate on reshaping the library as research collaborator. Current literature on academic library structures consists largely of institutional case studies (**Andrade & Zaghloul, 2010**; **Chan & Soong, 2011**; **Franklin, 2009**; **Larsen & Riis, 2012**; **Nesdill, Love & Hunt, 2010**; **Quinlan & McHarg, 2012**; **Renfro & Neal, 2012**; **Shore, 2012**; **Warren, 2011**), with few comprehensive surveys. In North America, ARL last surveyed library organization in 1996, although the organization and staffing of specific elements of research support have been examined more recently, including digital humanities, research data management, and scholarly communication (**Bryson, Posner, St. Pierre, & Varner, 2011**; **Eustis & Kenney, 1996**; **Fearon, Gunia, Pralle, Lake & Sallans, 2013**; **Radom, Feltner-Reichert, & Stringer-Stanback, 2012**). In the UK, the convergence and merging of university libraries with information technology and other academic support services has dominated the discussion of library structures over the past two decades (**Bulpitt, 2012**; **Field, 2001**; **Hanson, 2005**; **Pugh, 1997a**).

Few writers on library organization relate their discussion to classic organizational theory, notable exceptions being **Pugh's (1997b, 2005, 2007**) efforts to establish a theoretical basis for the management of technology-driven change in libraries and **McGuigan's (2012)** essay identifying theories relevant to managing change in academic libraries. Although library case studies rarely use terminology from organization design literature, it is evident that the dominant library model of recent decades is basically the mixed structure (**Child, 1988**), which organizes technical/back-office processes (e.g., acquisitions, cataloging) and generic front-of-house services (e.g., circulation) along functional lines and specialized client services in a divisional or market-based arrangement (i.e., individuals or teams devoted to supporting particular subjects, disciplines, departments, or schools). Many libraries have overlaid an element of matrix structure, by introducing cross-cutting roles to coordinate service provision across different groups (e.g., designating senior liaison librarians as lead specialist for key areas, such as scholarly communications); an emerging variant of this model uses functional specialists who do not have liaison assignments to supplement and support their liaisons (**Jaguszewski & Williams, 2013**).

1.3 Definition of key terms

Organization design has been defined as "the allocation of resources and people to a specified mission or purpose and the structuring of these resources to achieve the mission" (**Duncan, 1979**, p. 60). **Duncan (1979, p. 60)** also asserts that "Ideally, the organization is designed to fit its *environment* and to provide the information and *coordination* needed" (emphasis added). **Stanford (2013, p. 19)** offers an alternative definition of the concept emphasizing other dimensions:

"Arranging how to do the work necessary to effectively and efficiently achieve a business purpose and strategy whilst delivering high-quality customer and employee experience. Arranging involves aligning the organization with the strategy, creating coherent designs, while building trust among key stakeholders."

Design is thus conceived as the broader higher-level concept, with *structure*forming an important part of the design. However, the terms are often used synonymously, within both management and library literature. Another important point to note here is that "resources" in this context can include "tangible resources such as money and equipment and intangible resources such as knowledge" (**Miles, Snow, Fjeldstad, Miles, & Lettl, 2010, p. 93**).

1.4 Research question, aims and objectives

The central research question for the study is: *How are academic research libraries and their institutions organizing resources and services to support the research enterprise?*

The overall aims of the research are:

- To review organizational designs of research libraries and evaluate their fitness for purpose in the context of researcher support needs.

The specific objectives of the project are:

- To review trends and developments in library support and services for research and researchers
- To examine design dimensions and structural features of a sample of research library organizations
- To identify specific elements of library organization of particular relevance in supporting research
- To analyze strengths and weaknesses of potential organizational models.

2. METHODOLOGY

This research uses an exploratory sequential mixed methods design (**Creswell & Plano Clark, 2011**) to survey organization structures of research libraries in the specific context of library service responses to challenges presented by the contemporary e-research environment. The purpose was first to qualitatively explore dimensions of library structural designs with a small single-country sample and then to quantitatively determine the extent to which the qualitative findings generalized to a large international sample. The present paper reports the results from the first phase only, but describes the complete design to explain the rationale for the initial research.

In the first (instrument-development) phase, data were collected via institutional websites, organization charts, strategy documents, job descriptions and other public-domain sources from a purposive sample of 24 universities in the UK. The Russell Group of leading UK research universities was used as the sampling frame.1 The use of websites and web-based documents as sources of data and objects of analysis has been recognized as an appropriate method of investigation for contemporary researchers, particularly in relation to organizational research (**Bryman, 2004**). Previous studies have used this type of Internet-based research to investigate library strategy documents, either as the sole method of data collection (**Corrall, 2007**; **Pacios, 2004**), or as the first phase of a multi-method study (**Aldrich, 2007**; **Corrall, 2008**; **McNicol, 2005**).

Both deductive and inductive strategies were used in the initial phase. A review of related literature informed the systematic browsing and keyword searching of websites, but the key outputs of the research (such as the categorizations of structural arrangements) were generated from the data analysis. Evidence was captured and recorded electronically as raw data (document extracts or whole documents), notes and initial codes. The data were analyzed using matrix analysis techniques, using color coding to highlight examples with striking features, which were then re-examined to select cases illustrating sharp contrasts (**Corrall, 2007**). Data matrices are particularly suited to making large amounts of qualitative data accessible and meaningful, enabling insights to emerge, and supporting both cross-site and within-site comparisons (**Miles & Huberman, 1994**; **Nadin & Cassell, 2004**).

The findings from this initial exploration will be used to develop assessment measures that can be administered to a large sample. In the planned second phase, data on service configurations, reporting relationships, divisions and specializations, coordination and integration will be collected from research library directors in Australia, New Zealand, North America, and selected European countries, using an online questionnaire survey. The reason for collecting qualitative data initially was that there is little guiding theory on the subject to inform the development of a survey instrument (**Creswell & Plano Clark, 2011, p. 86**).

3. RESULTS

Formal organization structure charts were found via web searches in eight cases, i.e., for one-third of the sample investigated (*n*=24). However, relevant dimensions and salient features of the structures in use at other sites were identified by browsing their web pages and consulting staff directories, service descriptions and other information available (including structure charts for the institution in some cases, and further particulars for job announcements in a few instances). The main elements examined were service configurations, reporting lines, internal organization, specialist positions, and coordination mechanisms. Throughout the analysis, the focus was specifically on elements of library organization of particular relevance in supporting research.

The headline findings from the data are as follows:

- Integrated library and computing services have fallen from favour
- Reporting lines for library and information services have shifted
- Library structures continue to mix function and market/subject elements
- New senior positions and titles are signalling strategic priorities
- Subject liaison arrangements are being enhanced and remodelled
- Specialist research positions are emerging as a significant trend
- Dedicated spaces for researchers are improving service visibility
- Groups and websites are promoting boundary-spanning services.

3.1 Service configurations

One-third of the services featured some form of convergence between the library and one or more cognate professional services, most commonly the computing/information technology (IT) service, which often also includes learning/education technology. This finding contrasts with the position a decade ago when more than half of the 20 institutions then in the Russell Group were converged and one-quarter had merged structures with fully integrated library and IT provision for several core functions (e.g., help desk, training). Among the currently converged organizations, four had retained the same basic structure for a decade or more, of which one was still a merged service, two were converged, and one was federated (separate organizations within an information services division of professional services).

Two organizations had deconverged their library and IT services and combined the library with other cognate services to form an academic services directorate and a student services directorate respectively. Two other institutions had formed newly converged services in the last five years, in both cases bringing the library and IT together, into an information directorate in one case, and academic services in the other (which also included student services and quality enhancement). However, the key finding here is that most members of the group now have separate library organizations, which includes four high-profile services that have deconverged their merged or converged service organization within the last three years (King's College London, the London School of Economics (LSE), and the Universities of Bristol and Nottingham).

The variety of service configurations was a feature of UK converged information service organizations in the 1990s and 2000s, and this pattern has continued as services have de-converged and re-converged more recently. For example, at both the University of Birmingham and the University of Exeter, Library services is one division within an Academic Services directorate of seven divisions, but the composition of each directorate and structure of the library division varies. Table 1 shows the breakdown (based on data collected in summer 2013).

At divisional level, in addition to library services, common functions here include academic development/educational enhancement, student employability (careers services), and student services/registry functions (combined at Exeter). However, technology services (including administrative systems, academic computing and student IT services) are part of the converged service at Exeter, but at Birmingham, in the re-converged service, IT Services is a separate unit within Professional Services reporting to the Registrar & Secretary – unlike the former converged information service organization at Birmingham, where IT was a central part of the original unified structure. At the library divisional level, Table 1 shows that the primary split into four functional sections is similar in both institutions, namely: academic liaison/engagement; customer services/support;

Table 1. Sample composition of academic services directorates

University of Birmingham		University of Exeter	
Academic Services Divisions	*Library Services Departments*	*Academic Services Divisions*	*Library & Culture Services Groups*
• Birmingham Foundation Academy • Centre for Learning & Academic Development and Learning Spaces • Student Employability • Library Services • Special Collections • Registry • Student Services	• Academic Liaison, Learning, Research Support & Academic Skills • Collection Management & Development • Library Customer Support	• Academic Support Unit (Cornwall) • Education Quality & Enhancement • Employability & Graduate Development • Exeter IT • Library & Culture Services • Planning & Resources • Student Services	*Library & Research Support Teams* • Academic Engagement • E-Strategy and Resource Development • Heritage Collections • Library Customer Services *Arts & Culture Team*

collection/resource development; and special/heritage collections. However, Birmingham has departed from established practice by making Special Collections a separate division (not reporting to the Director of Library Services), while Exeter has brought in Arts and Culture as an additional service group reporting to the Assistant Director of Academic Services/Head of Library and Culture Services.

The make-up of the academic support department/team within each library also differs slightly. The Birmingham department includes responsibility for the University's Academic Skills Centre (which provides support in areas such as academic writing, maths and stats, and study skills), in addition to academic liaison, and research support (including supporting the REF and research data management); whereas Academic Skills Development at Exeter is located within the Education Enhancement division, and the Library's Academic Engagement group includes separate teams specializing in digital content and library marketing, and open access and data curation, in addition to liaison support for teaching and research. (At Birmingham, responsibility for managing and supporting the institutional repository and open access is part of the Collection Development and Management department.) Both institutions use the traditional subject librarian model to carry out liaison activities: at the time of data collection, Birmingham had 15 Subject Advisors supporting 5 Colleges, while Exeter had 10 Subject/Liaison Librarians supporting 4 Colleges and separate Business and Medical Schools (part of the College structure at Birmingham). The numbers of liaisons are commensurate with the respective sizes of the populations served, Birmingham being approximately 50 percent larger than Exeter.

The Information Services directorates that have continued as merged, converged or federated services are now more uniform in their service portfolios. All include libraries (and archives), IT (academic computing and corporate systems), and learning technologies, usually including responsibility for both equipping and managing teaching spaces, but not covering academic skills (other than information literacy), which are typically provided by academic support/learning development units, nor other student services functions. Within the sample, although some instances of shared support functions were found (e.g., Operations and Central Projects at Birmingham; Strategic Support, Finance, HR, etc. at Cardiff; Planning and Resources at Exeter), only Edinburgh currently displays elements of the more radical merged service structures that were adopted to varying extents by Birmingham and other leading research universities from the mid 1990 s.

The key features of the Edinburgh structure are the combined library, e-learning and IT first-line help/support service, and the organization of liaison librarians and IT professionals into three College-facing consultancy teams within the User Services Division. The consultancy teams are in a different part of the organization from other library staff, who are mainly located in the Library and Collections Division. This means they are separated from staff working in related research and learning support services, including teams and sections dealing with scholarly communication, and collection development and management, and also do not report to the *de facto* library service head, the Director of Library & University Collections.

3.2 Reporting relationships

In UK universities, the Librarian traditionally reported to the Vice-Chancellor or Principal (as the chief executive officer of the institution), which often differentiated the library from the IT/computing service; but the adoption of more formal management arrangements in higher education institutions over the past two decades and the reconfiguration of administrative support functions into groups of professional services has resulted in different structures. Nearly three-quarters of the library directors in the sample report directly (or indirectly, via another senior officer) to the University's Registrar/Secretary/Chief Operating Officer. Chief Operating Officer (COO) is becoming the dominant title in use, found in half of the institutions.

Only Oxford and Cambridge seem to have retained their traditional reporting line in its original form, although the head of the converged service at Edinburgh (who holds the formal title of Vice-Principal for Knowledge Management, CIO and University Librarian) reports directly to the Principal. At the time of data collection, the university librarian/library director at Manchester and Imperial reported to the equivalent of the Deputy Vice-Chancellor (Deputy Rector at Imperial); at Leeds and Nottingham, they reported to a Pro-Vice Chancellor (PVC), in both cases the PVC with responsibility for education/teaching and learning. The grouping of the library with education/student services is another notable trend in structural arrangements, reflected in the recently converged academic service model adopted at Exeter, the reconverged academic services directorate at Birmingham, and reconverged student services at Queen Mary (University of London). In addition, the deconverged library service at King's now reports to the College Secretary via the Director of Students & Education; similarly at Bristol, the reporting line is via the Deputy Registrar, responsible for "the student-facing and education-related services".

It is debatable whether a PVC reporting line is preferable to the COO line, particularly where the PVC portfolio has a focus on teaching/students. Reporting directly to the COO role can position the library director alongside other professional service heads, including both student services and research services, which is particularly valuable in the context of collaborative service development with the research office or equivalent in areas such as managing research data and measuring research impact.

3.3 Internal organization

The size and shape of the library superstructures vary significantly, with leadership teams ranging from three to eight members. The titles of senior management positions and labels used for the main divisions/departments/operational units also vary, and some libraries combine one or more functions together, but overall there was a striking consistency in the way activities were grouped into the following five areas:

- Academic services/research and learning support (generally subdivided by subject field)
- Collection development/information resource management
- Special collections and archives/heritage collections
- Customer services/reader services/user support
- E-strategy/digital services/information technologies.

The academic (learning and research) liaison/support function was explicitly represented in the senior team in three-quarters of the sample. In one-third of these (six cases), the relevant manager had other responsibilities (e.g., Planning & Resources, Collection Development/Management, Special Collections, User Services, Service Development). In one novel example, at the University of Manchester, research services, teaching and learning support, and academic engagement were distributed among three different members of the leadership team, and combined with other related responsibilities. At Edinburgh, the merged Information Services organization separates consultancy including library liaison services (in the User Services Division) from other library-based research and learning services (in the Library & Collections Division).

Academic Services was the descriptor used most consistently, found in the job titles of seven senior managers. However Research, or Research & Learning (or Learning & Research) featured in seven titles of associate/assistant directors/heads of division or equivalent. The word Liaison was included in four titles, and the term Academic Engagement in two jobs. Other variants included the position of Head of Relationship Management at York, and the five Heads of Faculty Services in the senior team at Southampton (where academic liaison is directed by the Deputy Librarian). The most complex title identified was Assistant Director – Academic Liaison, Learning & Research Support, and Academic Skills at the University of Birmingham (a position that has subsequently been changed to Academic Engagement).

The collection development/management function was equally prominent, also being represented in the senior teams of three-quarters of the sample. However, the term "collection" was used explicitly in slightly less than half the sample, and the titles of the relevant roles, divisions, or sections also varied within these cases, e.g., Collection & Access Management (Nottingham), Collection Development & Description (Cambridge), Collection Services (Durham,

Leeds). The term "resources" seems to be gaining currency as an alternative to "collection", suggesting the concept of the library collection may become less prominent as migration from purchased printed materials to licensed e-journals and e-books continues. At York, the relevant senior position is Head of Content.

In approximately half of the cases featuring a collection development or management role in the top team, the collections function was combined with the management or strategic development of library systems (in two cases along with facilities), represented by roles such as Head of Collections and eStrategy (Sheffield – a position that has now evolved into Assistant Director for Academic Services by adding responsibility for the Learning & Research Services team to the Procurement & Content and Digital Services & Systems teams), Head of eStrategy & Resource Development (Exeter – a position that is now combined with academic liaison responsibilities in an enlarged Academic Support portfolio), and Head of Information Management & Library Development (Bristol). A slightly less prevalent, but equally significant, trend (found in around one-quarter of the total sample), was the creation of senior roles solely concerned with the library's information systems/information technologies portfolio, notable examples here including: Associate Director, Digital Library Programmes & Information Technologies (Oxford), Head of Digital Library Services (Leeds), and Head of Digital Technologies & Services (Manchester).

Around half of the institutions had special collections and/or archives represented as a distinct role in the library's senior team at the level of divisional/departmental head, including two who used the alternative title of Head of Heritage Collections (Durham and Exeter). Also in around half of the sample, delivery of frontline services to users (such as lending/circulation and enquiry/help services) was represented explicitly in the senior group, e.g., Assistant Director – Customer Services (Kings), Head of Library Customer Services (Exeter). In several other cases, customer/user services was part of a larger portfolio for a member of the executive. Customer Services (or Customer Support) was by far the most common label used for this function, followed by four instances of User Services, two of Reader Services, and only one where the term Public Services was used.

3.4 Subject liaison

The importance of subject-based academic support was evident in the structures of all 24 libraries in the sample. With the exception of the Universities of Cambridge (where the library's strategic objectives include development of "integrated, subject-based provision for teaching and learning as well as research") and Manchester (where academic liaison through teams of faculty librarians was recently discontinued), every institution currently uses some form of subject librarian arrangement to provide specialized support to students and academic/research staff. Seven use the title *Subject* Librarian (or Subject Advisor or Subject Consultant) for this role, while 11 use *Liaison* Librarian (or Academic Liaison Librarian, or Faculty Liaison Librarian). (Other titles found include Academic Support Librarian, College Team/Support Librarian, Faculty Team Librarian, and Information Specialist.) In ten cases, the library had a team structure mirroring the institutional structure of Colleges or Faculties, with subject/liaison librarians led by Heads of Faculty Services, Faculty Team Leaders, Faculty Librarians, or similar. Some libraries had additional structural differentiation by including *Assistant* Liaison Librarians, or Liaison *Assistants* in their teams.

The strength of liaison teams/subject support varied across the sample from six liaison librarians at Queen Mary (University of London) and eight at Durham, to 32 subject librarians at Cardiff and 52 at Oxford (at the time of data collection). Although generally the liaison librarian title was associated with smaller-scale provision, this was not universally the case, as the Queen's (Belfast) structure included only eight Subject

Librarians (led by three Faculty Librarians), whereas Southampton had 24 Liaison Librarians, led by five Heads of Faculty Services (who also fulfilled liaison roles). The mean number across the sample was 15.9 and the median was 13.5. However, these figures should be seen as approximations only. They do not always represent the total number of librarians involved in subject-based learning support and research services, as many libraries have recently established specialist positions that focus solely on learning or research support, or a particular aspect of academic support. In addition, the method used to collect data on subject liaisons did not differentiate between full-time and part-time liaison staff.

3.5 Specialist positions

As reported in the Introduction, support for research has traditionally been a formal part of the subject/liaison/faculty librarian role, alongside support for teaching and learning, but published literature indicates a trend towards libraries putting more effort into the delivery and promotion of support for researchers, having identified opportunities to provide more specialized "higher-end" services, and also become aware that researchers often have a limited view of what librarians provide (tending to focus on resource procurement and collection management). A significant development in the past decade has thus been the creation of new roles within the academic services/support area designed to provide specialist support for researchers in areas such as literature searching, systematic reviews, current awareness, reference management, citation analysis, publishing advice, open access and data curation.

One-third of the sample libraries had established a specialist position with a research support focus, mostly within the past five years. In some examples, the role is similar to and positioned alongside existing subject/liaison librarians, with titles such as Academic Liaison Librarian (Researcher Support) at Durham, or Research Support Librarian at

Queen Mary. In most cases, the post-holder is also expected to fulfill a management coordination or strategic leadership role, both carrying out specialist work and providing direction for research-related service developments across library teams; examples of relevant job titles here include Research Support Leader at Liverpool, Library Liaison Manager (Research) at King's, Head of Scholarly Communications at Southampton, and Academic Support Manager (Research) at Warwick – where the post-holder is supported by a dedicated Academic Support Librarian. In a few instances, the post-holders also act as subject librarians/academic liaison contacts (e.g., Senior Consultant, Research Support at Cardiff, and Research Support Services Manager at LSE).

In addition to this type of cross-cutting coordination and development role, several libraries have established more specific research support positions, whose remits have evolved in line with additional needs identified. For example, the Research Skills & Development Librarian at Cambridge directs the University Library's Research Skills Programme, but is also a member of the University's Open Access Project team; at Imperial, the intended focus of the new position of Research Support Librarian (Medicine) was mediated literature searching for systematic reviews, but the role developed to provide support in other areas, such as data management. Other longer established examples of specialist research support provision in libraries include two largely grant-funded units, the Support Unit for Research Evidence (SURE, established in 2000) at Cardiff, which specializes in systematic reviews of the literature; and the Centre for Research Communications at Nottingham (established formally in 2009, but with a history going back to the SHERPA project of 2002–2006), which carries out research and development work in scholarly communication, open access and related areas.

Management of the institutional repository/research archive is now an established library function, but its location within the organization structure varies. In two instances within the sample, the repository function had been subsumed within the institutional research information system managed by the research office; in another two examples, the repository service was a joint venture with the university computing service, but located in the library. However, repository management was mostly a library responsibility, generally located either in the library's systems/IT/digital services division (10 examples), or with academic/research and learning services (9 examples). In two cases the research support librarian or equivalent was formally designated as the institutional repository manager, and in another two cases, the research support leader had formerly filled the role of repository manager. At Birmingham, the repository is the responsibility of the Collection Development and Management department, where the Digital Assets team is located (in the absence of a library IT department).

Policy developments and funding mandates related to research assessment, open access and data management have resulted in many new and redesigned specialist library positions and teams to manage and coordinate institutional support in these areas. For example, Edinburgh has a Scholarly Communications Team of four people providing support for policy development and implementation, open access advice and advocacy, intellectual property rights, bibliometrics, repository administration, publications submission, and metadata quality. Exeter has an Open Access and Data Curation Team, comprising a Manager, a Data Curation Officer, and an Open Access Administrator. At Bristol, an institutional Research Data Service was launched in summer 2013, as a Library-led collaboration with IT Services and Research & Enterprise Development; it is staffed by a Manager, two Senior Research Data Librarians and an Assistant Research Data Librarian. Sheffield has formed a new Research Support Team within its Learning & Research Services group, which includes two Coordinators for Open Access and Research Data Management, two Open Access Assistants and a Research Data Management Liaison Assistant.

3.6 Dedicated spaces

The University of Warwick Library launched its Wolfson Research Exchange in 2008 as an innovative suite of specialist services and dedicated technology-rich spaces for researchers, with provision for quiet study, interdisciplinary networking, academic meetings, and social activities, and a programme of events on topics such as funding opportunities, grant applications, data management, journal impact, and spinout companies, "offered not only by the library but by other university departments supporting research too" (**Carroll, 2011b, p. 55**). Novel services include Research Match ("a unique service which profiles your research and makes it visible") and Collaborative Sandpits (interactive multi-day workshops to create project proposals for funding).

With separate programming for graduate research students, early career researchers, and academics/research staff, the facility has its own dedicated staffing, with a Research Exchange and PG Hub Manager, a Coordinator, Support Officers, and Assistants, who work alongside other managers and academic support specialists in the Academic Services division of the Library. *In situ*support is provided by a team of part-time Advisers, who are all current or recent PhD students (**Carroll, 2011a**). The Library website presents the facility as "A space, a website, a community", reflecting its several dimensions, including its hosting of networking evenings, interdisciplinary special interest groups, online peer to peer discussions, and blogs.

The development of dedicated purpose-designed space for researchers emerges as a significant trend creating physical presence for renewed efforts by libraries to position themselves as research partners and collaborators. The new research spaces typically have card-controlled entry to restrict access and demonstrate that libraries understand that researchers want and need physical facilities that are separate from undergraduates. Queen Mary opened a new 58-seater Research Reading Room for graduate researchers and academic staff in 2011, after a consultation identified the need for a quiet, comfortable working environment away from busier noisier parts of the Library. The facility was "designed along similar lines to The British Library Reading Rooms, with high specification study desks finished in oak and leather, providing larger than standard worktop space, and LED task lighting at each desk". Birmingham, similarly, has a Researchers' Room "reserved for use by postgraduate research students and academics only".

Social space and technology provision generally differentiate contemporary researcher facilities from traditional reading rooms. For example, Durham is developing a researchers' study area for the exclusive use of postgraduate research students, with individual, collaborative and social areas, which is also being used to promote information about library support for research. At Exeter, the Old Library housing the University's special collections has been refurbished and renamed the Research Commons, with a new large reading room, enhanced IT, multi-media seminar room, and café-style break out space and terrace area. At York, where the Library was able to extend into an adjoining building, renovation has created new improved research space, including a study area, specifically for researchers; an informal research lounge area; and – the most unusual marketing feature – a Research Hotel, a suite of bookable rooms that visiting academic staff or researchers can reserve for their exclusive use as temporary work areas. The website explains that "The hotel concept does not extend to the provision of sleeping arrangements, but does offer use of the research lounge".

3.7 Boundary-Spanning Activities

Policy developments in the UK, notably the Engineering and Physical Sciences Research Council Policy Framework on Research Data (**EPSRC, 2011**) and the Research Councils UK (**RCUK, 2013**) Policy on Open Access have given rise to other structural arrangements, such as steering/working groups bringing institutional stakeholders together to facilitate compliance with funder requirements. Such groups are typically led or chaired by a Pro-Vice Chancellor for Research or senior research administrator and usually include representatives of academic schools/faculties/colleges in addition to key staff from the library, IT service and research office or equivalent. For example, at Cambridge, an Open Access Working Group was set up in 2012 to draft an institutional policy and make recommendations on implementation, followed by an Open Access Project Board in 2013 "to manage the transition to Open Access". At Exeter, "an Open Access and Research Data Management Policy Task and Finish Group was set up in March 2012 to develop an institutional policy on research data and Open Access issues". Leeds, University College London (UCL) and Warwick all have a two-tier structure for managing research data, with a high-level steering/oversight group or project board complemented by a Research Data Working Group at Leeds, a Technical & Operations Group at UCL, and an Action/Implementation Group at Warwick; the UCL structure also includes a User Group, which meets less frequently.

Institutional websites offer another opportunity to coordinate, focus and promote services to support research and researchers. Three-quarters of the library and information services in the sample had a set of web pages bringing together information targeted at researchers. At one-third of the sites, there was a prominent link from the library/information services home page, labeled Research Support, Researchers, or simply Research; at another eight, users could reach the relevant page with two clicks. The web pages varied in the topics featured, but there was a clear trend towards selectivity and integration, bringing together resources judged most useful to researchers, offered both by the library and other institutional units (such as research computing/IT services and the research office), and also trying to present the information from a research user viewpoint, rather than a service provider perspective. For example, at King's College London, the Library Research Support home page has the sub-title "Support through the research lifecycle" and a choice of six categories to click: Ideas, Research funding, Managing information, Disseminating research, Evaluating research, and Preserving research; there is also a link to the Library Research Support site from the institutional Research Support website. The Researcher@Library website at Leeds "gives researchers at the University of Leeds a single point of access to information, support and resources across the University".

Research impact, open access, and data management feature strongly on these web pages, reflecting current preoccupations of the UK research community, and also demonstrating how libraries are taking advantage of recent policy developments to position themselves as central to the research process. In almost all institutions in the sample, the library has been assigned responsibility for managing the funds allocated for Article Processing Charges incurred in compliance with funder requirements for gold open access. Most libraries have developed new web pages to support this process, typically including tables and/or decision trees/flow charts adapted to show how the institution is interpreting and implementing the relevant policies. Web pages supporting research data management were found at two-thirds of the institutions, with half of the sites clearly branded as part of the library/information services offer. Five services (including two projects led by the library) were presented as university-level offerings, usually acknowledging contributions by different services, and two services were based in the IT service. The websites were at various stages of development, with two consisting solely of links to external sites.

4. DISCUSSION AND CONCLUSIONS

4.1 Convergence

The move away from the converged service model is one of the striking findings of the study. However, the distinction between "organizational" and "operational" convergence (**Field, 2001**, p. 268) is important here. *Organizational*convergence (i.e., formally combining two or more services to form one organization) is evidently in decline within the sample, especially structural designs merging libraries with computing/IT services. *Operational* convergence (i.e., separate services/departments collaborating to coordinate their activities to improve conference and effectiveness), in contrast, is arguably more prevalent than ever, with libraries extending and deepening their collaborations and partnerships beyond IT and educational development colleagues to other professional services, such as research offices. This is particularly evident in:

- services (e.g., institutional repositories, research data services) delivered as collaborative ventures with IT services and/or research offices
- relationships and structures (e.g., committees and groups) supporting boundary-spanning activities in areas such as data management and open access
- unified research support websites providing a single point of access to university-wide resources for researchers, which could be described as an example of a *virtually* converged information service.

4.2 Positioning

The location of the library/information service organization in the institutional organizational structure is rarely discussed in the literature, but is important because it has implications for the allocation of resources, the coordination of activities, and the perception of the service by stakeholders. The two significant trends here were first, the administrative positioning of the library within professional services reporting to the Chief Operating Officer or equivalent, which was identified as the dominant model; and secondly, an increasing tendency to group libraries with "student-facing and education-related services" through reporting lines and directorate structures. The latter is arguably not a positive move for libraries trying to reposition themselves as credible collaborators and partners in research activities.

4.3 Specialization

The survey showed that departmentalization within libraries has evolved beyond the traditional reader/technical services model to a focus on five strategic areas of activity: information resources, academic engagement, customer service, heritage collections, and digital technologies, although there was significant variation in the size of leadership teams. Use of the term "Research" in the job titles of around one-third of the senior management positions identified sends a signal to stakeholders about the priority attached to this area. The strong focus on information systems/technologies identified at the senior management level is not surprising, but it was interesting to find that in many cases the technology portfolio was combined with another area of responsibility, and most frequently with information resources/collections, which in effect continued the tradition of locating systems librarians in a technical services division/department.

Specialization of library professional staff continues to focus on the subjects/academic disciplines taught and researched in the institution, but subject/liaison roles are increasingly being complemented and supplemented by coordinating roles (such as research support managers) and functional specialists for particular areas of activity (e.g., research data management, open access publishing). The hybrid approach to specialization found in UK research libraries echoes structural developments reported in North America (**Covert-Vail & Collard, 2012**; **Jaguszewski & Williams, 2013**).

An interesting dimension of specialization is the growing trend to extend the development of specialized research support into dedicated spaces for researchers. The space-as-service concept is gaining currency internationally as a visible contribution to research, particularly in support of early career researchers, and interdisciplinary collaboration, with the research commons beginning to emerge as a parallel development to the successful learning commons movement (**Corrall & Lester, 2013**; **Daniels, Darch, & de Jager, 2010**;**Ohio State University Libraries, 2012**). UK examples range from traditional reading rooms to the fully-serviced technology-rich space at Warwick, using early career researchers as peer support advisors, and offering novel services such as research profile matching and suites of bookable rooms for individuals and meetings (**Carroll, 2011a,b**).

4.4 Coordination

Coordination within libraries is being formalized through roles such as research support managers and leaders, and data management and open access coordinators (and also by internal committees and working groups). Coordination with other services and operations is being pursued through and facilitated by institution-wide groups at both strategic and operational level, typically established to deal with external mandates, but offering the potential of additional benefits for librarians in relationship building, professional credibility, and institutional positioning. As noted above, websites are an additional valuable way of coordinating research services to provide a single point of access for researchers and can be used not only to pull together library and other institutional resources and services, but again to position the library as a major player in the research arena. Research support websites complement the work of research support coordinators in libraries, but can also fulfill that role in libraries which do not have an individual tasked with that responsibility.

4.5 Outcomes

The purpose of the present study was to explore qualitatively how a sample of academic research libraries are organizing resources and services to support their research mission with the intention of identifying important dimensions of organizational design that could be used to develop an assessment instrument for a larger scale international survey. The study achieved its objectives and the results both confirm and extend the findings of related research investigating library plans, roles and skills for supporting and serving researchers. Although the sample was limited to 24 libraries in one country, the results identify structural variables related to library support for research that can be used as a provisional framework for libraries in the UK and other countries to examine and evaluate their own structures.

The present study was conceived as the first phase of a larger scale international survey of organizational design in research libraries and the findings are being used to develop a questionnaire to collect quantitative data on important structural and contextual variables identified from this exploratory study and ongoing review of related literature. The dimensions to be investigated include:

- positioning of libraries within their parent institutions
- scope of library/information service organizations
- size and focus of library leadership teams
- specialization of units, roles and space
- formal and informal convergence of activities
- mechanisms facilitating boundary-spanning activities.

NOTES

[1]http://www.russellgroup.ac.uk.

ACKNOWLEDGMENTS

A summary of this study was presented at the 2013 LIBER Annual Conference in Munich, Germany. A synopsis of the key findings was also published in *Library Connect Newsletter, 11*(3), The Library and the Research Enterprise (pp. 3–4).

REFERENCES

1. Adema, J., & Schmidt, B. (2010). From service providers to content producers: New opportunities for libraries in collaborative open access book publishing. New Review of Academic Librarianship, 16(sup. 1), 28–43.

2. Aldrich, A.W. (2007). Following the phosphorous trail of research library mission statements into present and future harbors. ACRL 13th National Conference, March 29–April 1, 2007, Baltimore, MD (pp. 304–316). Chicago, IL: Association of College & Research Libraries. Retrieved April 30, 2014

3. Andrade, R., & Zaghloul, R. (2010). Restructuring liaison librarian teams at the University of Arizona Libraries, 2007–2009. New Library World, 111(7/8), 273–286.

4. Auckland, M. (2012). Re-skilling for research: An investigation into the role and skills of subject and liaison librarians required to effectively support the evolving information needs of researchers. London: RLUK Research Libraries UK.

5. Bourg, C., Coleman, R., & Erway, R. (2009). Support for the research process: An academic library manifesto. Dublin, OH: OCLC Research.

6. Bradbury, K., & Weightman, A. (2010). Research support at Cardiff University Library. SCONUL Focus, 50, 65–70.

7. Brunner, M.L. (2010). PhD holders in the academic library: The CLIR Postdoctoral Fellowship Program. In K. Williams & S. Walter (Eds.), The expert library: Staffing, sustaining, and advancing the academic library in the 21st century (pp. 158–189). Chicago, IL: Association of College & Research Libraries.

8. Bryman, A. (2004). Social research methods. (2nd. ed.). Oxford: Oxford University Press

9. Bryson, T., Posner, M., St. Pierre, A., & Varner, S. (2011). Digital humanities: SPEC Kit 326. Washington, DC: Association of Research Libraries

10. Bulpitt, G. (Ed.) (2012). Leading the student experience: Super-convergence of organisation, structure and business processes. London: Leadership Foundation for Higher Education. Retrieved April 30, 2014, from http://www.lfhe.ac.uk/en/components/publication.cfm/S3%20-%2005

11. Carlson, J., & Kneale, R. (2011). Embedded librarianship in the research context: Navigating new waters. College & Research Libraries News, 72(3), 167–170.

12. Carroll, D.L. (2011a). Fostering a community of scholars at the University of Warwick: The Wolfson Research Exchange. New Review of Academic Librarianship, 17(1), 78–95.

13. Carroll, D. (2011b). What researchers want: A personal case study. SCONUL Focus, 51, 52–56.

14. Chan, D.L.H., & Soong, S.C. (2011). Strategic repositioning in a dynamic environment. Library Management, 32(1/2), 22–36.

15. Child, J. (1988). Organization: A guide to problems and practice (2nd ed.). London: Paul Chapman

16. Corrall, S. (2007). Benchmarking strategic engagement with information literacy in higher education: Towards a working model. Information Research, 12(4), paper 328.

17. Corrall, S. (2008). Information literacy strategy development in higher education: An exploratory study. International Journal of Information Management, 28(1), 26–37.

18. Corrall, S., & Lester, R. (2013). The researcher's view: Context is critical. In L. Watson (Ed.), Better library and learning spaces: Projects, trends and ideas (pp. 183–192). London: Facet

19. Corrall, S., Kennan, M.A., & Afzal, W. (2013). Bibliometrics and research data management services: Emerging trends in library support for research. Library Trends, 61(3), 636–674.

20. Cotta-Schønberg, M. (2007). The changing role of the subject specialist. Liber Quarterly, 17(3/4). Retrieved April 30, 2014

21. Covert-Vail, L., & Collard, S. (2012). New roles for new times: Research library services for graduate students. Washington, DC: Association of Research Libraries. Retrieved April 30, 2014, from http://www.arl.org/storage/documents/publications/nrnt-grad-roles-20dec12.pdf

22. Cox, A., Verbaan, E., & Sen, B. (2012). Upskilling liaison librarians for research data management. Ariadne, 70. Retrieved April 30, 2014, from http://www.ariadne.ac.uk/issue70/cox-et-al

23. Creamer, A.T., Morales, M.E., Kafel, D., Crespo, J., & Martin, E.R. (2012). A sample of research data curation and management courses. Journal of eScience Librarianship, 1(2), 88–96.

24. Creswell, J.W., & Plano Clark, V.L. (2011). Designing and conducting mixed methods research (2nd ed.). Thousand Oaks, CA: SAGE

25. Crow, R., Ivins, O., Mower, A., Nesdill, D, Newton, M., Speer, J., & Watkinson, C. (2012). Library publishing services: Strategies for success. Washington, DC: SPARC. Retrieved April 30, 2014, from http://wp.sparc.arl.org/lps/

26. Daniels, W., Darch, C., & de Jager, K. (2010). The research commons: a new creature in the library. Performance Measurement and Metrics, 11(2), 116–130.

27. Delserone, L.M., Kelly, J.A., & Kempf, J.L. (2010). Connecting researchers with funding opportunities: A joint effort of the Libraries and the University Research Office. Collaborative Librarianship, 2(1), 33–37.

28. Drummond, R., & Wartho, R. (2009). RIMS: The Research Impact Measurement Service at the University of New South Wales. Australian Academic & Research Libraries, 40(2), 76–87.

29. Duncan, R. (1979). What is the right organization structure? Decision tree analysis provides the answer. Organizational Dynamics, 7(3), 59–80

30. Duranceau, E.F. (2008). The "Wealth of Networks" and institutional repositories: MIT, DSpace, and the future of the scholarly commons. Library Trends, 57(2), 244–261.

31. EPSRC (2011). EPSRC policy framework on research data. Swindon: Engineering and Physical Sciences Research Council. Retrieved April 30, 2014, from http://www.epsrc.ac.uk/about/standards/researchdata/Pages/policyframework.aspx

32. Eustis, J.D., & Kenney, D.J. (1996). Library reorganization & restructuring: A SPEC kit. Washington, DC: Association of Research Libraries

33. Fearon, D., Gunia, B., Pralle, B.E., Lake, S., & Sallans, A.L. (2013). Research data management services: SPEC Kit 334. Washington, DC: Association of Research Libraries

34. Federer, L. (2013). The librarian as research informationist: A case study. Journal of the Medical Library Association, 101(4), 298–302.

35. Field, C.D. (2001). Theory and practice: Reflections on convergence in United Kingdom universities. LIBER Quarterly, 11(3), 267–289.

36. Franklin, B. (2009). Aligning library strategy and structure with the campus academic plan: A case study. Journal of Library Administration, 49(5), 495–505.

37. Fonseca, A.J., & Viator, V.P. (2009). Escaping the island of lost faculty: Collaboration as a means of visibility. Collaborative Librarianship, 1(3), 81–90. Retrieved April 30, 2014

38. Furlough, M.J. (2010). The publisher in the library. In S. Walter & K. Williams (Eds.). The expert library: Staffing, sustaining, and advancing the academic library in the 21st century (pp. 190–233). Chicago, IL: Association of College and Research Libraries. Retrieved June 23, 2014

39. García-Milian, R., Norton, H.F., Auten, B., Davis, V.I., Holmes, K.L., Johnson, M., & Tennant, M.R. (2013). Librarians as part of cross-disciplinary, multi-institutional team projects: Experiences from the VIVO collaboration. Science & Technology Libraries, 32(2), 160–175.

40. Giesecke, J. (2010). Finding the right metaphor: Restructuring, realigning, and repackaging today's research libraries. Journal of Library Administration, 51(1), 54–65.

41. Gold, A. (2007). Cyberinfrastructure, data, and libraries, Part 2. Libraries and the data challenge: Roles and actions for libraries. D-Lib Magazine, 13(9/10). Retrieved April 30, 2014

42. Greyson, D., & Surette, S. (2013). "You're just one of the group when you're embedded": report from a mixed-method investigation of the research-embedded health librarian experience. Journal of the Medical Library Association, 101(4), 287–297.

43. Hanson, T. (Ed.) (2005). Managing academic support services in university: The convergence experience. London: Facet

44. Harris-Pierce, R.L., & Liu, Y.Q. (2012). Is data curation education at library and information science schools in North America adequate? New Library World, 113(11/12), 598–613.

45. Healy, A.M. (2010). Increasing the visibility of the library within the academic research enterprise. Issues in Science and Technology Librarianship, 63.

46. Herther, N.K. (2009). Research evaluation and citation analysis: key issues and implications. The Electronic Library, 27(3), 361–375.

47. Housewright, R., Schonfeld, R.C., & Wulfson, K. (2013a). Ithaka S+R US faculty survey 2012. New York: Ithaka S+R. Retrieved April 30, 2014, from http://www.sr.ithaka.org/research-publications/us-faculty-survey-2012

48. Housewright, R., Schonfeld, R.C., & Wulfson, K. (2013b). Ithaka S+R/JISC/RLUK US survey of academics 2012. New York: Ithaka S+R. Retrieved April 30, 2014,

49. Jahnke, L.M., & Asher, A. (2012). The problem of data: Data management and curation practices among university researchers. In L. Jahnke, A. Asher & S.D.C. Keralis, The problem of data (pp. 3–31). Washington, DC: Council on Library and Information Resources.

50. Jaguszewski, J.M., & Williams, K. (2013). New roles for new times: Transforming liaison roles in research libraries. Washington, DC: Association of Research Libraries. Retrieved April 30, 2014

51. Larsen, S. & Riis, C. (2012). Aarhus University Library: Restructuring a university's library organisation. Bibliothek Forschung und Praxis, 36(2), 235–240. 0030.pdf

52. Lee, C.A., & Tibbo, H. (2011). Where's the archivist in digital curation: Exploring the possibilities through a matrix of knowledge and skills. Archivaria, 72, 123–168.

53. Long, M.P., & Schonfeld, R.C. (2014). Ithaka S+R Library Survey 2013. New York: Ithaka.

54. Lynch, C. (2008). The institutional challenges of cyberinfrastructure and e-research. EDUCAUSE Review, 43(6), 74–88.

55. Lyon, L. (2012). The informatics transform: Re-engineering libraries for the data decade. International Journal of Digital Curation, 7(1), 126–138.

56. MacColl, J., & Jubb, M. (2011). Supporting research: Environments, administration and libraries. Dublin, OH: OCLC Research.

57. Manchester University Library (2012). News from member libraries. SCONUL Focus,56, 61–62.

58. McBain, I., Culshaw, H., & Walkley Hall, L. (2013). Establishing a culture of research practice in an academic library: an Australian case study. Library Management, 34(6/7), 448–461.

59. McGuigan, G.S. (2012). Addressing change in academic libraries: A review of classical organizational theory and implications for academic libraries. Library Philosophy and Practice, Paper 755.

60. McNicol, S. (2005). The challenges of strategic planning in academic libraries. New Library World, 106(11/12), 496–509.

61. Miles, M.B. & Huberman, A.M. (1994). Qualitative data analysis: An expanded sourcebook. (2nd. ed.). Thousand Oaks, CA: SAGE

62. Miles, R.E., Snow, C.C., Fjeldstad, O.D., Miles, G., & Lettl, C. (2010). Designing organizations to meet 21st-century opportunities and challenges. Organizational Dynamics, 39(2), 93–103.

63. Nadin, S. & Cassell, C. (2004). Using data matrices. In C. Cassell & G. Symon (Eds.), Essential guide to qualitative methods in organizational research. (pp. 271–287). London: SAGE.

64. Nesdill, D., Love, A., & Hunt, M. (2010). From subject selectors to college and interdisciplinary teams. Science & Technology Libraries, 29(4), 307–314.

65. Nowviskie, B. (2013). Skunks in the library: A path to production for scholarly R&D. Journal of Library Administration, 53(1), 53–66.

66. Nutefall, J.E., & Chadwell, F.A. (2012). Preparing for the 21st century: Academic library realignment. New Library World, 113(3/4), 162–173.

67. Ohio State University Libraries (2012). Overview of research commons in academic libraries: A white paper. Columbus, OH: Ohio State University Libraries. Retrieved April 30, 2014,

68. Pacios, A.R. (2004). Strategic plans and long-range plans: Is there a difference? Library Management, 25(6/7), 259–269. doi: 10.1108/01435120410547913. Retrieved June 23, 2014

69. Posner, M. (2013). No half measures: Overcoming common challenges to doing digital humanities in the library. Journal of Library Administration, 53(1), 43–52.

70. Pugh, L. (1997a). Convergence in academic support services. London: British Library Research and Innovation Centre

71. Pugh, L. (1997b). Some theoretical bases of convergence. New Review of Academic Librarianship, 3(1), 49–66.

72. Pugh, L. (2005). The management of hybrid libraries. Library and Information Research, 29(92), 13–31. Retrieved April 30, 2014,

73. Pugh, L. (2007). Change management in information services (2nd ed.). Abingdon: Ashgate

74. Quinlan, C., & McHarg, H. (2012). The emerging library: Structure, culture, and lessons learned from the dissolution of a combined libraries–IT organization. Journal of Library Administration, 52(2), 147–161.

75. Radom, R., Feltner-Reichert, M., & Stringer-Stanback, K. (2012). Organization of scholarly communication services: SPEC Kit 332. Washington, DC: Association of Research Libraries

76. RCUK (2013). RCUK policy on open access and supporting guidance. Swindon: Research Councils UK. Retrieved April 30, 2014

77. Renfro, P., & Neal, J.G. (2012). The integration of libraries and academic computing at Columbia: New opportunities for internal and external collaboration. Journal of Library Administration, 52(2), 162–171.

78. Schrader, A., Shiri, A., & Williamson, V. (2012). Assessment of the research learning needs of the University of Saskatchewan librarians: A case study. College & Research Libraries, 73(2), 147–163.

79. Shore, E. (2012). Embracing hybridity: The merged organization, Alt/ac and higher education. Journal of Library Administration, 52(2), 189–202.

80. Stanford, N. (2013). Organization design: Engaging with change (2nd ed.). Abingdon: Routledge

81. Stanton, J.M., Kim, Y., Oakleaf, M., Lankes, R.D., Gandel, P., Cogburn, D., & Liddy, E.D. (2011). Education for eScience professionals: Job analysis, curriculum guidance, and program considerations. Journal of Education for Library and Information Science, 52(2), 79–94

82. Vandegrift, M., & Varner, S. (2013). Evolving in common: Creating mutually supportive relationships between libraries and the digital humanities. Journal of Library Administration, 53(1), 67–78.

83. Vaughan, K.T.L., Hayes, B.E., Lerner, R.C., McElfresh, K.R., Pavlech, L., Romito, D., ..., Morris, E.N. (2013). Development of the research lifecycle model for library services. Journal of the Medical Library Association, 101(4), 310–314.

84. Vinopal, J., & McCormick, M. (2013). Supporting digital scholarship in research libraries: Scalability and sustainability. Journal of Library Administration, 53(1), 27–42. Retrieved June 23, 2014

85. Warren, V. (2011). Using workforce structures to enable staff retention and development: An academic library case study. New Library World, 112(1–2), 8–18.

86. Yakel, E., Conway, P., Hedstrom, M., & Wallace, D. (2011). Digital curation for digital natives. Journal of Education for Library and Information Science, 52(1), 23–31. Retrieved June 23, 2014.

87. Young, H., & Lund, P. (2008). Reflections on a benchmarking survey of research support provided by 1994 Group libraries. SCONUL Focus, 43, 51–56.

4

CHAPTER

The e-Reader — an Educational or an Entertainment Tool? e-Readers in an Academic Setting

Peter Ahlroos, Jonna Hahto

Tritonia Academic Library, Vaasa, Finland

ABSTRACT

In this paper the authors will discuss a pilot project conducted at the Tritonia Academic Library, Vaasa, in Finland, from September 2010 until May 2011. The project was designed to investigate the application of e-readers in academic settings and to learn how teachers and students experience the use of e-readers in academic education. Four groups of students and one group of teachers used Kindle readers for varied periods of time in different courses. The course material and the textbooks were downloaded on the e-readers. The feedback from the participants was collected through questionnaires and teacher interviews. The results suggest that the e-reader is a future tool for learning, though some features need to be improved before e-readers can really enable efficient learning and researching.

Keywords: e-readers, higher education, academic libraries

INTRODUCTION

Aim

In the autumn of 2010 The Tritonia Academic Library started an e-reader project with funding allocated by the city of Vaasa, in honour of the city's 400-year anniversary. In 2010 there was a growing interest in e-readers and tablet computers, especially in the local press, which also spurred the idea to start a project to examine the use of portable reading devices in an academic setting. The purpose was also to gather more knowledge about e-books and e-readers in libraries and to promote the services of Tritonia.

The aim of the project is to investigate the application of e-readers in an academic setting and to learn how teachers and students experience the use of e-readers in academic education.

About Tritonia

Tritonia is a joint academic library for five universities: the University of Vaasa, VAMK University of Applied Sciences, Åbo Akademi University, Novia University of Applied Sciences and Hanken School of Economics. Tritonia also provides pedagogical services for the staff of all five universities. Of the 12,000 students, 70% have Finnish as their mother tongue, while the remaining 20% are Swedish-speaking.

The printed collections consisting of 336,000 volumes are available for all students and staff of the five universities and universities of applied sciences, as well as for external library users. By contrast, all electronic resources are licensed for a specific organisation and only the students and staff of the particular organisation have access to those. Because of this, one single e-book may have to be acquired separately for all the five organisations. This is challenging for both the library staff and the library users.

In addition to traditional library services, Tritonia also offers educational support at the EduLab unit. The activities that Edulab is based upon are pedagogical courses (25 ECTS), guidance and media production, which are offered for free to the staff of all five universities. Since the start in 2001, the main goal has been staff development and support.

BACKGROUND

The e-book and e-reader market is booming at the moment. Amazon already announced that they are selling more Kindle books than print books and Apple's iPad dominates the tablet computer market (Amazon.com, 19.4.2011). In Europe the e-book market has evolved at a slower pace, mainly because of the lack of content and the high price of e-readers and e-books, but the Kindle's arrival to Europe has livened up the market.

Libraries have also invested in e-books, have purchased e-book databases, have tried out lending e-readers to customers and set up pilot projects. Even so, despite of their best efforts, the libraries have not really forged their way into the e-book business. However, for the future it is more vital than ever that libraries continue to experiment with different e-book models and e-book and e-reader solutions.

Research in the Same field

Several studies concentrating on e-books and articles about e-readers were published at the beginning of the 21st century. However, the devices never really broke through until the past couple of years. At the moment more and more studies again focus on e-readers, e.g., Clark et al., (2008) dealt with user experiences and users' opinions about e-readers in their survey conducted at Texas A & M University Libraries. The participants in the study, members of the library and university staff, found the Kindle reader interesting, but questioned the potential usefulness of the device in research libraries on account of licensing issues, graphic display and cost.

Pattuelli and Rabina (2010), on the other hand, attempted to understand the effects of reading devices on everyday reading practices and also potential applications for library services. They gathered feedback from 20 library and information science students, who had used the devices for one week. The results were mixed; however, the usability issues were outweighed by the device's portability and readers' convenience of use anywhere.

Both Clark et al., (2008) and Pattuelli and Rabina (2010) focused mostly on leisure reading. The use of e-readers in an academic setting and the potential application of the e-reader as an educational tool have been much less explored themes ine-reader studies. There are a few exceptions, including pilot studies conducted by seven universities in the USA (Damast, 2010) and Aalto University in Finland (Aaltonen et al., 2010). In both studies the students were given e-readers with course material for a certain period of time. The studies encountered certain usability issues. For example, students felt that the lack of colours and the difficulties in navigation, browsing, bookmarking and taking notes made the devices awkward to use.

An Academic Setting — a Challenge to e-Readers

The purposes for reading and the methods used can offer a starting point for a discussion of the challenges for e-readers, because the way we read varies with the type of literature we read (see Marshall, 2010, p. 33). For example, reading a fiction novel differs substantially from reading a textbook, as in leisure reading the text usually advances linearly. Reading for study and research purposes involves a lot of browsing, glancing, seeking and re-reading. Annotation and navigation are important ways to interact with the text (Marshall, 2010, pp. 37–36.). In order for e-readers to be considered a real alternative or replacement for paper books and computers, reading devices should be able to support different ways of using the devices.

Today's reading devices are mainly focused on linear reading or, more specifically, reading for entertainment. Most of the devices on the market have annotation and search functions, but as studies indicate, users often find these functions awkward and slow to use (Aaltonen et al., 2010; Damast, 2010). Also, browsing books is often found difficult. Another possible weakness emphasized in an academic setting is the lack of colours, as pictures, charts and tables are important parts of scientific documents.

The different formats can also create difficulties for students and researchers. e-Books in ePUB format, for example, do not have static page numbers, which makes it very hard to provide precise page numbers for citations. PDFs do not have the same problem, but because the PDF format is not scalable, books and documents in this format may require scrolling and are therefore challenging to read on some e-readers (Ballhaus et al., 2010, p. 3).

TEACHING, LEARNING AND E-BOOKS

The significance of information technology in teaching cannot be underrated as the students are already familiar with information technology both in informal and formal settings. They are part of the Generation Y, the Net Generation, accustomed to navigate in the constant information flow. However, the filtering strategies they use to cope with large quantities of information can often be described as hasty and superficial, which reflects how they regard and use information (Parment, 2008). This generation is also more demanding than the previous one, expecting teachers and librarians to keep up with the new technology.

As Cox (2010) points out, the teacher's attitudes and beliefs as well as the teacher's willingness to adopt new teaching methods and practices affect the perceived value of IT in learning. Unfortunately, studies indicate that once their training is finished, teachers rarely see the need to question or change their teaching practices, including those related to the use of IT (Cox, 2010, pp. 16–17) A number of researchers have looked at the influence of new technologies on student achievement and learning processes, but the results are both for and against (Minčić-Obradović, 2011, pp. 174–175). For one thing, the potential change tends to be temporary and superficial and teaching and learning methods are seldom transformed profoundly.

e-Books have the potential of being an efficient and effective educational tool, as the combination of text and multimedia can enhance readers' understanding of abstract concepts and theories (Ballhaus et al., 2010, p. 13; Collier and Berg, 2011, p. 32; Minčić-Obradović, 2011, p. 161). Features such as customisability of text size, the ability to convert text to audio and a built-in encyclopedia also have benefits in education (Minčić-Obradović, 2011, p. 161). Textbooks read on e-readers enable real-time dialogue and discussion between teachers and students, with options for sharing notes, comments and questions. Many e-readers do not yet have these features, but technical development has been so rapid that no doubt we get to see those in the future.

METHODOLOGY

The project was designed to investigate the application of e-readers in an academic setting and to learn how teachers and students make use of e-readers during courses. The data was mainly collected through questionnaires and interviews. In the next paragraphs, we will discuss the choice of e-reader, preparing the test groups, and the data collection methods.

The main reasons for choosing Amazon's Kindle reader were: a user-friendly interface, an affordable price, a range of functions such as dictionary and text-to-speech and the possibility to download one e-book onto several devices. The closed environment, the lack of EPUB support and the lack of Finnish or Swedish content were not considered major issues, as the aim of the project was not to concentrate on usability issues or to evaluate the device itself, but to focus on the pedagogical aspect of e-readers in general.

The e-reader project started in September 2010, once the third-generation Kindle became available, and ended in May 2011. During the project five different groups of students experimented with e-readers. The groups are presented in Figure 1.

Subject and course	Number of students in the course	Period of use	Books
Technical Communication, research course in advanced studies	10	4 months	George F. Hayhoe: A Research Primer for Technical Communication
Journalistik, Mass media and communication	9	4 weeks	Joseph Turow: Media Today (3rd Edition)
Information Technology, database management in advanced professional studies	6	4 weeks	M. Lisin, J. Joseph and A. Goyal: Microsoft SQL Server 2008 Reporting Services Unleashed
Language (Swedish), Digital texts	11	4 weeks	Books from project Gutenberg, Maria Turtschaninoff's fantasynovel 'Underfors'
Learning in Networks 1, teacher group	15	3 weeks	Confronting the Challenges of Participatory Culture, Living and Learning with New Media

Figure 1. Test groups.

The groups chosen for the project were small, advanced courses, where the teacher was able to choose the literature used in the course. We approached specific teachers because we knew them to be interested in trying out new methods and techniques. The books and other documents chosen by the teachers were downloaded on 25 Kindle readers (wifi). Later on five additional Kindles were acquired. Kindle DX, Sony PRS-650 reader touch edition, Bokeen Cybook OPUS and iPad I were also acquired, mainly for demonstration purposes.

A short introduction (20–30 minutes) was arranged for the test groups and the students were asked to sign a lending agreement. The participants were also given the possibility to download their own documents on the Kindle reader. The introduction was the only contact with the students and further communication was handled by the teacher.

The students were asked to answer online questionnaires before and after the test period. Due to the project timing and resourcing, the only manageable way of collecting data was through a questionnaire. The questionnaires included open questions as well as multiple choice questions. In the first questionnaire the respondents were asked about their technical skills, studying habits and attitudes towards new technology in general. The second questionnaire related to studying in the course, the use of the device and respondents' thoughts about e-readers, studying and learning. Three teachers were interviewed to get in-depth insight into students' reactions and teachers' thoughts about e-reader applications. We also had access to a blog article and six reviews written by the students as additional assignments in their course.

In addition to the test groups, e-reader and e-book theme days were organised. During these three days over 200 students, staff members and other visitors acquainted themselves with different e-readers and the library's e-book collections. 67 visitors also filled in a questionnaire, which contained questions about the library's e-book collections and attitudes towards e-readers.

RESULTS

89% of the students filled in the first questionnaire but only 56% (20 of 32) of the students answered the second questionnaire, despite of several reminders sent by email. In the teacher group 10 of 15 teachers filled in the questionnaire. Both the students and the teacher group can be categorised as early adopters or as an early majority on the innovation adoption curve of Rogers (2003, 281), as shown in the graph below. The majority of the respondents estimated their technical skills above average.

In the next sections the respondents' attitudes towards the Kindle reader and Kindle's pedagogical value will be discussed.

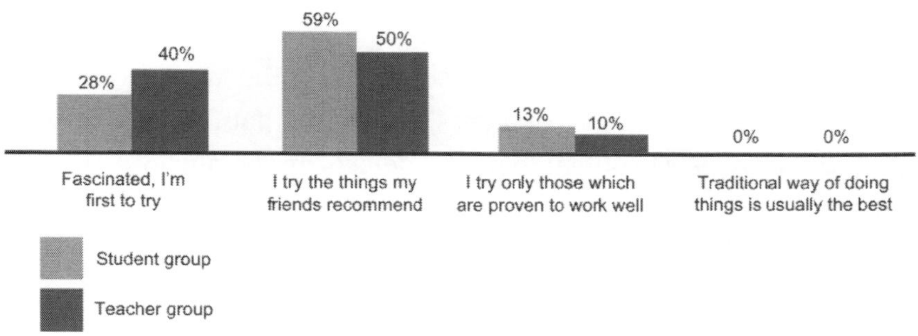

Figure 2. Respondents' attitude towards new technologies.

Respondents' Attitudes towards the Kindle Reader

One of the findings of the theme days and the student group meetings was that the e-readers seem to lack the wow-factor, which makes tablet computers like the iPad so attractive. As a couple of respondents pointed out, the lack of colours, basic button navigation and poor internet browsing capabilities made them think of 'old black-and-white

TVs' and 'the Middle Ages'. Still, half of the respondents said they would rather read documents on a basic e-reader than on a computer screen.

The teacher group appeared to be slightly more critical towards the Kindle reader than the students, especially judging by the answers to the open questions. In many comments the Kindle reader was compared to the iPad and found to be inadequate, as 'Kindle cannot compete with iPad, there's so much more you can do with an iPad'. Even though the students in general gave lower values for different features, they seemed to be more neutral in their attitudes towards the Kindle reader. Many students did not seem to care whether they read the book in paper format or on an e-reader as long as they had access to the book. Based on the students' comments one may speculate that changing the reading habits and 'getting used to the different format', as several students put it, is really a bigger issue for the students than the shortcomings in usability.

The teacher interviews also gave clues about the way the students experienced the use of the e-readers, indicating that for them it would not necessarily be a problem to have a device for one purpose only. The students seem to want to keep their studies and free time apart, and for this reason it is natural for many of them to have different devices for different purposes. That is, the iPad is for 'all the fun stuff' like surfing, whereas the Kindle reader is a more serious device, meant for studying and reading. One of the students actually noted that it is a good thing that the Kindle reader doesnot have any unnecessary functions, because it is easier to concentrate on studying.

One of the teachers observed that the students seemed to take the course more seriously and that they were more active and ambitious than usual, but it is impossible to determine if this was a reaction to the new, interesting technology or if there simply were more enthusiastic students in the group than usual.

In Figure 3, the overall user experience is shown. Nearly every respondent had a positive or at least a neutral experience with the Kindle reader. A majority of both the teachers and the students also estimated that the effect of the e-reader on studying and learning on the course was neutral or slightly positive.

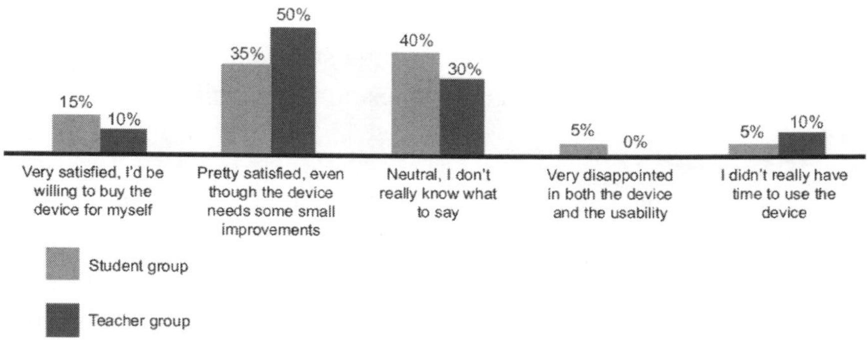

Figure 3. Overall user experience.

As assumed by previous studies (Aaltonen et al., 2010; Ballhaus et al., 2010;Damast, 2010), the features the respondents missed the most were a touch screen, colours, internet browsing, real page numbers and scandic letters. Many of the teachers were especially dissatisfied with Kindle's PDF support, as reading the PDF file requires a lot of navigation and the size of the text is small. The most appreciated features were the readability of the text, the portability, and the long battery life of an e-reader.

The Kindle Reader in Learning and Teaching

Only a few respondents had previous experience with e-readers, so they — as was expected — found it difficult to evaluate the potential application of the device in teaching and learning. The students were most pleased that they did not need to chase textbooks in the library and carry heavy book bags to school. The three teacher interviews gave some insight into possible ways of using e-readers.

Teacher 1 found it very important that his students, as future journalists, would read and explore text in the same way as their future readers will explore the text written by the students. The use of the e-reader caused the teacher to rethink his teaching methods and also to present a concrete plan as to how the course could be developed utilising the Kindle reader to enhance the students' understanding of the subject. He also had ideas about using the Kindle for presenting reports.

Teacher 2 held this course for the first time and therefore did not yet have any ideas for renewing it. The idea of the course was to explore digital texts and the teacher got special permission from the publisher to use a Swedish fantasy novel as course material. The students were satisfied that they could read a book without others knowing what they read and that they could alter the font size. As future primary school teachers, they also speculated whether the e-reader could increase the motivation for reading among boys. These students felt that reading fiction would enhance the experience with multimedia and linking to web pages. Especially the fantasy novel with many references to real places would have given them a better reading experience if they could have used the web while reading.

The third interviewed teacher was not fully convinced of the benefits of e-readers in learning and teaching, mainly because of the difficulties in browsing books and the poor support for PDF files. In her courses books are usually used as reference literature and with the Kindle it was hard to navigate between chapters and get an overview of the book. She also uses lots of PDF articles, which are awkward to read on the Kindle since they require scrolling.

In her opinion, the ideal teaching aid would not be the traditional book, but a service where a selection of different devices would be offered together with digital materials and pedagogical support. Another function she desired was a shared notation system.

DISCUSSION AND CONCLUSIONS

Even though the study was a pilot project and the results may not be generalised because of the small and homogeneous sample, the feedback has been encouraging and we intend to continue the project in the autumn. There are features such as browsing, PDF support and an internet connection, that need to improve before e-readers can enable efficient learning and researching in an academic setting. However, though many of the respondents still preferred the paper book, nearly all of them saw the e-reader as a future tool for studying.

As expected, those teachers with sparse experience of either the Kindle, the content of a specific e-book or the course, had few ideas on how to develop their courses. One month of e-reader usage seems to be too short a time period and the teacher test period should really start much earlier than the course. Consequently, in the future the project will focus on teachers and they will be offered a longer loan period.

The vision for the future of libraries is that students and teachers would be able to download library material easily from a distance on their own devices. For the time being e-books or e-readers are not a solution, especially because the vast majority of the libraries' present e-book collections are not compatible with most of the e-readers, and the books read on e-readers are often available only for private customers. Logistic issues such as charging the batteries and preparing the e-readers also require a great deal of time and effort from the library staff, if the reading devices are lent to library patrons. However, if libraries want to become part of the development and the chain of distributing e-books, it is essential that they experiment with various e-book distribution models and e-readers.

The volume of e-resources increases every year and remote access to resources reduces the need to visit the library. This does not necessarily mean that libraries are less important than before, it merely represents a change in patrons' needs. As the collections are more and more in digital format and the self-service increases, libraries have more resources to concentrate on the quality of service rather than the quantity. There is a greater need for information services, personal guidance and information literacy education than ever.

Finally, from our point of view the project can be considered a success merely because it got some of the teachers to rethink their methods of teaching and ways to develop the course. As a whole, the project shows that academic libraries can contribute to teaching methods by providing tools and support to teachers to develop their teaching methods and acquire new ideas and technology.

REFERENCES

1. Aaltonen, M., P. Mannonen, S. Nieminen and M. Hjelt (2010): 'A collaborative study: on the demands of mobile technology on virtual collection development', World Library and Information Congress: 76th IFLA General Conference and Assembly. 10–15 August 2010, Gothenburg, Sweden.

2. Amazon.com, news release 19.4.2011: 'Amazon.com Now Selling More Kindle Books Than Print Books'.

3. Ballhaus, W., M. van der Donk and P. Stokes (2010): 'Turning the pages: the future of ebooks', Technology, Media & Telecommunications, pwc. http://www.pwc.com /gx/en/entertainment-media/publications/future-of-ebooks.jhtml.

4. Clark, D. T., S. P. Goodwin, T. Samuelson, and C. Coker (2008): 'A qualitative assessment of the Kindle e-book reader: results from initial focus groups'. Performance Measurement and Metrics, 9(2), 118–129.

5. Collier, J. and S. Berg (2011): Student learning and e-books. In: S. Polanka (ed.), No shelf required: e-books in libraries (pp. 19–36). Chicago: American Library Association.

6. Cox, M. (2010): The changing nature of researching information technology in education. In: A. McDougall (ed.), Researching IT in education: theory, practice and future directions (pp. 11–24). New York, NY: Routledge.

7. Damast, A. (2010): 'E-Book Readers Bomb on College Campuses'.BusinessWeek (June 11, 2010), 3.

8. Marshall, C. C. (2010): Reading and Writing the Electronic Book. Synthesis Lectures on Information Concepts, Retrieval, and Services #9, San Rafael, California: Morgan & Claypool Publishers.

9. Minčić-Obradović, K. (2011): E-books in academic libraries. Oxford, UK: Chandos.

10. Parment, A. (2008): Generation Y: framtidens konsumenter och medarbetare gör entré. Malmö: Liber.

11. Pattuelli, C. M. and D. Rabina (2010): 'Forms, effects, function: LIS students' attitudes towards portable e-book readers'. Aslib Proceedings62(3), 228–244.

12. Rogers, E. M. (2003): Diffusion of innovations. 5th ed. New York: Free Press.

5

CHAPTER

European Research Libraries and Special Collections

Graham Jefcoate

University Library, Radboud University Nijmegen, P.O. Box 9100, 6500 HANijmegen, The Netherlands.

ABSTRACT

An interesting, and perhaps rather unexpected, aspect of current thinking on the future of research libraries has been a new emphasis placed on the role of rare books and special collections. As the delivery of traditional library services, for example collection building and description, becomes more dependent on national or regional consortia, so libraries are considering what will remain to be done at the local and institutional level. Various library "trend watchers" have suggested local activities on which individual libraries may wish to concentrate in the future, including information literacy, customised user support in information retrieval, and so on. A surprising number have also mentioned "legacy" collections of printed material and other special collections as a likely focus of future professional efforts at the local library level.

INTRODUCTION

An interesting, and perhaps rather unexpected, aspect of current thinking on the future of research libraries has been a new emphasis placed on the role of rare books and special collections. As the delivery of traditional library services, for example collection building and description, becomes more dependent on national or regional consortia, so libraries are considering what will remain to be done at the local and institutional level. Various library "trend watchers" have suggested local activities on which individual libraries may wish to concentrate in the future, including information literacy, customised user support in information retrieval, and so on. A surprising number have also mentioned "legacy" collections of printed material and other special collections as a likely focus of future professional efforts at the local library level.

One aspect of this new emphasis is clearly the perceived need to digitise large quantities of print retrospectively, but the importance of locally-held special collections as resources for teaching and research has also been stressed. Special collections, each with unique materials or assemblages of materials, are seen as "critical identifiers" of individual research libraries. The period is drawing to a close during which the curatorship of rare books and special collections was apparently regarded (at least by some!) as a rather marginal activity within the research library.

Much of the current thinking about the future of special collections within the 21st century research library comes from the English-speaking world, and especially from the United States. The term "critical identifiers", for example, was introduced into discussions by a task force on special collections set up some years ago by the

Association of Research Libraries (ARL), the umbrella body for research libraries in the United States. In a number of countries on this side of the Atlantic there has been some concern about the need to open up a debate on the issue of special collections in European research libraries and to link up with current thinking in the English-speaking world. During 2003, for example, the Anna Amalia Bibliothek in

Weimar and the Initiative Fortbildung für wissenschaftliche Spezialbibliotheken und verwandte Einrichtungen e.V. invited the author to take part in the planning of a conference on the future of special collections, a conference that took place in Weimar in the autumn of 2005.

The Weimar conference, " Zwischen Kulturmanagement und Forschung: Sondersammlungen im 21. Jahrhundert" attracted a wide range of speakers and participants from the United States, the United Kingdom, The Netherlands, and of course from Germany itself, but also (and perhaps rather unexpectedly) from a number of other European countries. Sessions covered collection development, collection description, digitisation and collection care, but also fundraising and the promotion of collections. There was clearly widespread interest in the topics discussed and indeed enthusiasm for coordinated action. A consensus emerged in discussions at the conference that participants should remain in contact and that a number of follow-up activities should be organised. These have included so far:

- the publication of the Weimar papers [2];
- a study-tour to the United States by a group of German special collections librarians in summer 2006;
- a workshop in Berlin led by US experts in autumn 2006[3];
- and finally, the consideration of a possible "statement of commitment" to special collections by European research libraries.

One of the most pressing topics at the Weimar and Berlin workshops was the need to promote the significance of special collections more effectively, in other words, to consider how they could be moved up the agenda of parent institutions and sponsoring bodies. Again, the work of the ARL in the United States was seen as a possible model. The ARL

Special Collections Statement of Principles on "Research libraries and the Commitment to Special Collections" (2003), for example, was regarded by many as a de facto benchmark for special collections policy and practice in North America. Many of those involved in the Weimar workshop, felt that this "benchmark" - suitably adapted for the European situation - might well be a useful tool on this side of the Atlantic, for example in making a case for resources to governing bodies.

Liber as forum for discussion and action

In this context, LIBER was mentioned as a possible forum for discussion and action. With the support and encouragement of the LIBER Board, an informal working group with members from Germany, The Netherlands and the UK has drafted a statement from a European perspective. The text was drawn up by the author (Graham Jefcoate, Nijmegen), in cooperation with Michael Knoche and Jürgen Weber (Weimar) and David Pearson (London), and with significant contributions from Jan Bos (The Hague), Jutta Weber (Berlin), Bettina Wagner and Claudia Fabian (Munich), Dorothea Sommer (Halle) and others. Special mention needs to be made of the support and advice of our two US-based experts throughout the process, Alice Schreyer (Director, Special Collections Center, University Library, Chicago) and Alice Prochaska (Director, Yale University Library). The importance of their contribution can hardly be overstated.

LIBER members were invited to endorse the draft statement, which had the support of the LIBER Executive Board, at the Annual General

Assembly in Warsaw on Friday, 6 July 2007. The Assembly accepted the proposal unanimously. The statement will be included on LIBER's new website and LIBER is planning to issue a press release to ensure maximum publicity for the statement.

Meanwhile, consideration of the future role of special collections in the future research library continues. The LIBER Board has asked Ivan Boserup, Keeper of Manuscripts and Rare Books at the Royal Library in Copenhagen, to lead a small working party to consider the current representation of special collections-related issues within LIBER and to make recommendations to the Board by the time of the next LIBER conference in Istanbul in 2008. A number of colleagues will also be keeping in touch with developments in the United States. The present author, for example, has been asked with Richard Ovenden of the Bodleian Library (and one of the speakers at Weimar) to become a "corresponding member" of a new task force on special collections.

Liber statement of principles: european research libraries and their commitment to special collections [4]

Most academic and research libraries in Europe hold collections of rare and special materials, which may be variously defined. These encompass a wide range of documentary formats and other media, including historic and modern books, manuscripts and archives, music and maps, ephemera, photographs and sound recordings, and digital archives. As primary sources, or significant accumulations of materials relating to particular topics, they are essential for research across a range of disciplines. They have artistic, historic or research importance beyond their purely textual content that justifies their preservation as artefacts whatever surrogates may be available. They provide important evidence for our material, intellectual and cultural heritage and reflect our human diversity.

Members of the Ligue des Bibliothèques Européennes de Recherche (LIBER), in common with other academic and research libraries internationally, recognise the value and the particular obligations that the stewardship of these special collections represents. We collect, preserve and provide access to them on behalf of our parent institutions and the needs of international learning and scholarship. We regard these collections and our commitment to them as crucial to the fulfilment of our institutional mission.

In recognition of the obligations associated with special collections, members of LIBER therefore commit to the following principles:

- To ensure that the crucial importance of special collections to the missions of our libraries is clearly stated and understood;
- To continue to build special collections, including new collections where appropriate, in line with institutional collection development policies and other commitments;
- To understand and respond to the needs of users; to consult and collaborate with scholars and other user communities wherever appropriate;
- To provide appropriate levels of funding for the proper maintenance and preservation of special collections and for enabling the widest possible access to them for users;
- To make as much information as possible about our special collections accessible online, following established guidelines;
- To ensure the proper security of special collections and to ensure they are stored in conditions that are environmentally sound, following established guidelines;
- To include special collections in overall strategic planning and library development;
- To work collaboratively with appropriate partners on matters relating to collection building, preservation and improving access,

including the digitisation of materials and the provision of metadata to appropriate standards;

- To stimulate contacts with other memory institutions (including museums and archives) in order to exchange ideas, develop best practice and promote practical cooperation in the interest of users;
- To ensure that awareness of the significance of special collections, and the competencies needed to work with them, are preserved and enhanced among young professionals, and especially by offering training opportunities and support wherever possible;
- To inform stakeholders in parent institutions and the wider communities we serve about the obligations represented by the stewardship of special collections.

NOTES

[1] For further information about the contents of this article or the issues raised by it, please contact: Graham Jefcoate, Director General, University Library, Radboud University Nijmegen, P.O. Box 9100, 6500 HA Nijmegen, The Netherlands. Telephone: +31 (0)24-36 12400; email: graham.jefcoate@ubn.ru.nl

[2] Due to be published by Harrassowitz Verlag in autumn 2007.

[3] Moving special collections to the centre: a workshop for special collections administrators and curators".

[4] This statement of principles for the management of special collections, prepared after consultation with colleagues in European research libraries by a small virtual task force, led by Mr Graham Jefcoate, Director, Nijmegen University Library, was unanimously adopted by

LIBER members at the LIBER Annual General Assembly in Warsaw on Friday 6 July 2007.

WEBSITES REFERRED TO IN THE TEXT

1. ARL - Association of Research Libraries, task force on special collections. http://www.arl.org/rtl/speccoll/

2. ARL, Special Collections: Statement of Principles. http://www.arl.org/rtl/speccoll/speccollprinciples.shtml

3. Initiative Fortbildung für wissenschaftliche Spezialbibliotheken und verwandte Einrichtungen e.V. (Initiative for Continuing Education in Academic and Research-Oriented Special Libraries

   and Related Institutions). http://www.initiativef ortbildung.de/

4. Zwischen Kulturmanagement und Forschung: Sondersam mlungen im 21. Jahrhundert. Organisation, Dienstleistungen, Ressourcen. (Between Cultural Management and Research: Special Collections in the 21st Century. Organisation,Services,Resources). http://www. Initiativefort bildung. de/pdf/2005/ sondersammlungen_dt.pdf

6
CHAPTER

Consortium Negotiations with Publishers - Past and Future

Pierre Carbone

Conservateur général, Directeur du Service commun de la documentation Université Paris 12 et Coordonnateur consortium et bureau professionnel COUPERIN, Val-de-Marne 61, avenue du Général-de-Gaulle, 94010 CRETEIL cedex, France

ABSTRACT

Since the mid nineties, with the development of online access to information (journals, databases, e-books), libraries strengthened their cooperation. They set up consortia at different levels around the world, generally with the support of the public authorities, for negotiating collectively with the publishers and information providers general agreements for access to these resources. This cooperation has been reinforced at the international level with the exchange of experiences and the debates in the ICOLC seminars and statements. So did the French consortium Couperin, which is now gathering more than 200 academic and research institutions. The level of access and downloading from these resources is growing with geometrical progression, and reaches a scale with no comparison to ILL or access to printed documents, but the costs did not reduce and the libraries budgets did not increase. At first, agreements with the major journal publishers were based on cross-access, and evolved rapidly to the access at a large bundle of titles in the so-called Big deal. After experiencing the advantages of the Big deal, the libraries are now more sensitive to the limits and lack of flexibility and to cost-effectiveness. These Big deals were based on a model where online access fee is built on the cost of print subscriptions, and the problem for the consortia and for the publishers is now to evolve from this print plus online model to an e-only model, no more based on the historical amount of the print subscriptions, to a new deal. In many European countries, VAT legislation is an obstacle to e-only, and this problem must be discussed at the European level. This change to e-only takes place at a moment where changes in the scientific publishing world are important (mergers of publishing houses, growth of research and of scientific publishing in the developing countries, open access and open archives movement). The transition to e-only leads also the library consortia to deal with issues as preservation of print and electronic materials and perennial access to information.

INTRODUCTION

Library consortia appeared about fifty years ago. They promoted for decades sharing of resources and cooperation in collection development policies, they set up online union catalogues and interlibrary loans networks. So, it is natural they applied to share the access to online resources.

From these times, there is a continuous tension: scientific publications are a very little part (1 to 2 %) of the research costs (Kiley, 2007), but they are essential for research development and for innovative processes. So, two trends are coexisting: the continuous growth of scientific publications, and the barriers to access the whole publications.

At the beginning of the nineties, with internet, the roles of each of the partners in the publishing process changed. The authors could publish and give direct access to articles to their readers - such as did the Physics community in the early nineties with ArXiv.[1] The readers could directly make comments or suggest changes in articles and contribute to new versions. New publishing sectors emerged with the open access journals and the institutional repositories, that are now in competition with the commercial publishers. Libraries were facing a new challenge due to technological change: in the paper economics, providing materials and giving access to information are two distinct processes, in the online economics the two processes are merged. This revolution, that has not yet ended, gives a bigger place to the new materials (online journals, databases, e-books, ...) but it is not yet the end of paper materials, and the provision of online resources generate for the libraries budgets new costs which do not substitute to the former costs.

CONSORTIA – PARTNERS OF PUBLISHERS AND VENDORS

In the online information world, the consortial cooperation between libraries is more and more needed for a wide access to scientific publications. It is because the libraries united themselves in consortia that the publishers and the information providers recognized them as essential partners. Indeed the publishers find their interest in negotiating with consortia who are expressing clearly the needs of a community, and it is more easier for them than negotiating with single libraries, so they fostered this process allowing discounts proportional to the financial weight of a consortium and to the number of libraries involved. But the library consortia are not the agents of the publishers, their strength is in answering the needs of their members and of their users. Their action is of public interest, and is generally supported by the public authorities.

The existing consortia established an international cooperation and set up between 1997 and 2000 the International Coalition of Library Consortia (ICOLC). ICOLC is "an informal, self-organized group" comprising now 200 consortia from around the world - including developing countries. Within ICOLC, the consortia exchange their knowledge and learn from each other the best practices. If they discuss in these meetings with publishers and information providers on their respective policies and business models, and even if the big publishers are worldwide companies, ICOLC does not negotiate with vendors, but publishes statements and documents that are of interest for all the partners, and these statements are useful for applying in negotiations some rules accepted by all the partners.

In most cases, consortia negotiate directly with the vendors, but some consortia commit an external negotiator who is paid with a percentage of the discount he obtains from the vendor. So does JISC with Content complete Ltd.[2] Anyway, a good negotiation is a one that satisfies as much as possible the needs and interests of all the members and that is concluded by an agreement with the vendor. Of course, librarians must acquire new skills for this new job, but they are the most qualified for leading these discussions with vendors. They are the best for negotiating all the aspects of a licence agreement, not only price, but also content, access and services provided. They are also able to give a feedback from the users, and to suggest to the publishers actions for improving the quality of the service provided. For this reason, some publishers set up librarians advisory boards for exchanges on strategic and marketing issues.

But this partnership is not so equal between the two types of partners, especially for the online journals. If the access to most of the online databases may be obtained from several aggregators, and there is place in this sector for competition between vendors, the great journals publishing companies do not consent competition and they are the exclusive providers of their online services especially in the academic world. So the consortia have little space in negotiating the prices, and for the most they reached only agreements on moderation of the price increase of subscriptions (6 or 7% instead of 10% and more at the beginning of the year 2000). This moderation is aimed at guaranteeing some stability to the market and is a counterpart for allowing access to all the journals of the publisher (or if not, to a great number of titles). This is the so-called Big Deal.

BIG DEALS – PRINT AND ONLINE MODEL

The Big Deal was not the first business model for the consortia. At first, they obtained agreements for cross-access to the titles of a publisher that were subscribed at least by one member. The cross-access model had for consequence iterative renegotiations each time the members replaced twofold or manifold subscriptions by new titles to give access to more titles or each time a new member joined the consortium. This way, as consortia accepted progressively new members, they were keeping the choice of the content, but this model was not enough stable for the publishers, who rapidly proposed competitive conditions for the unlimited access to large bundles of journals, in Big Deals. The conditions were to agree with multi-year contracts, with "reasonable" increases per year (inferior to the previous annual increases), and with maintaining the amount of subscriptions (or cancelling titles for a marginal cost, in the limit of 1% of the total amount of the subscriptions).

With these Big Deals, the consortia gave access to their users to a number of online titles that could not be imagined before, and this for a reasonable increase of the budget. This extra cost has been generally supported at the beginning totally or partially by the funding authorities. The level of access and downloading knew a geometrical progression, and reaches now a scale with no comparison to ILL or to access to printed documents. So the Big Deals boosted the consortia, and the users - mostly the researchers - were accustomed in a way it is no more reasonable to imagine to go back to the past situation. More, they rapidly accessed to many titles that the libraries did not subscribe before, and they gather now 25% more articles in twice as many journals (European Commision, 2006)

For instance, the French consortium COUPERIN counted a few members in 1999 and expanded rapidly with the first big deals, and reaches now 210 members: universities, other academic institutions and research organizations ... As the 2005 figures show, if Couperin concluded 91 licence agreements with vendors, the activity of the Couperin members is concentrated on a few publishers: 46 % of the members contracted with 1 to 5 publishers; only 5 major publishers were in contract with at least 50 Couperin members (American Chemical Society - 53, Encyclopedia Universalis - 51, Elsevier - 123, Springer - 62, and Wiley -53), and these 5 publishers counted for half of the online costs paid by the members of the consortium (paper subscriptions excluded). This is in relation with the usage statististics. The number of articles downloaded from the six greatest publishers (American Chemical Society, Blackwell, Elsevier, Institute of Physics, Kluwer, Springer, Wiley) evolved from nearly 3 million in 2001 to nearly 12 million (of which 10 million from Science Direct) in 2005.

But the costs did not reduce and the income did not increase, the libraries could not cancel subscriptions for titles included in these Big Deals (or only in the margin of 1% of the turn-over). So the libraries cancelled subscriptions for titles from medium and little publishers (mostly learned publishers), the leadership of the great publishers was strengthened and had for consequence new mergers in these five last years. The result is a relative stabilization of the market for the great publishers, who have some margins of flexibility (for instance, they make changes in the list of the journals in the bundle). On the other part, there is not really flexibility for the libraries. The usage statistics show that generally 80% of the downloads are concentrated on 30% of the titles (Kohl , 2007), but the publishers don't accept to reconfigure these Big Deals and to reduce the turn-over. So, when libraries argue for a more restrictive bundle, only with the titles more accessed, the publishers don't accept to contract at a lower cost. So, cost-effectiveness issues are more and more important.

COST-EFFECTIVENESS

In front of this situation, some consortia experienced a 'crisis' mode of negotiation. At the beginning of a negotiation, nobody can say if it will be concluded by an agreement or not, and some hard discussions may include 'rupture' phases. But it is a failure if a 'rupture' position does not lead the partner to change his position. For instance, in 2004, Cornell and other universities in the USA cancelled their subscriptions with Elsevier, and so did Switzerland. But the users were the first punished and did not understand they would no more have access, and these consortia were obliged to come to a new agreement.

Ohio Link experienced another way, that was called an orderly retreat from the Big Deal (Gatten & Sanville, 2002), and that is an evolving Big Deal, the "border warfare model" (Kohl, 2007) based on two principles: "a reasonable increase in Big Deal is OK (it might be the inflation in the 3% range at present), if the inflation exceeds an acceptable amount, the consortium selects titles to cancel to reduce costs." This is possible if the consortium is able to analyze detailed use data and to reach agreement of his members on the titles to cancel, and has set up an information system. "In OhioLink's case in 2006 only 2 out of 8 publishers up for the Big Deal renewal/inflation increases decided for the border warfare model rather than acceptable inflation".

This solution works because it is a collaborative one. It is vital for the consortia to find such collaborative solutions agreed by their members and that allows them to reach positive agreements with the publishers. If they don't succeed, they are exposed to tensions and failures that endanger the life of the consortium itself. In a consortium are united different libraries, some are bigger, some are smaller, and the immediate interests of each of them could be opposite. The collaboration within a consortium is a force not only for the smaller, but for all. The strength of the consortium is in the link, it is in the network, in the ability of all its members to support the same decision, and not in the individual libraries. We all need this strength to conclude positive agreements with publishers.

MULTI-CONSORTIAL AGREEMENTS

To reduce the pressure of the great publishers and to maintain a large access to the others, consortia would also unite themselves. Some good news are showing this is possible. For instance, the Southern European Libraries Link (SELL) concluded a multi-consortial agreement with the Association of Learned and Professional Society Publishers (ALPSP). In the same way, Knowledge Exchange published a licensing tender on the European Union website for agreements applicable to United Kingdom, Denmark, Germany and The Netherlands. Such multi-consortial licensing tenders could work for supporting medium and little publishers, and also for accessing to some special resources which are needed only by a few libraries in each country and don't reach a critical mass in a single consortium.

But these are exceptions for the moment. All this new economics is based on ancient models, on the historical amount of paper subscriptions: the e-access fee is a percentage of this amount as the e-only fee is calculated with a discount on this amount. The publishers feared that with online access the libraries would unsubscribe paper journals, and guaranteed this way they would maintain (and annually increase) their income. Now, many libraries are willing to shift to e-only, but the e-only discount is not sufficiently incentive and in Europe the VAT legislations are a barrier to this evolution, as the VAT on electronic services is 10 to 18 % more expensive than the VAT on journals (SOFI, 2006). But a great part of the paper journals received in the libraries are no more used by their readers, and the costs in staff and premises could be spared if only a few libraries took the responsibility to preserve paper journals.

A new Big Deal between publishers and consortia adapted to the online publishing economics and making easier the shift to e-only has become a necessity. Libraries are not able to face annual increases such way, and publishers don't have interest in pushing the consortia to breaking-down. The publishers would move their positions, as they are confronted with two trends. First, they go through a new competition with researchers, the world of scientific publications is continuously evolving, and now there is a social consciousness that the results of publicly founded research would be in open access.[3] Open access journals and institutional repositories are taking more and more place. If many researchers imagine that these alternative models based on the author pay model could replace the reader pay model, for at least the twenty years coming now, it is more plausible that a new and moving balance will be established between these three sectors: commercial publishing that will keep the more important part, open access journals, institutional repositories. The second trend is more hopeful for the publishers, as scientific publishing is also enlarging and now it doesn't have borders in the world; new scientific nations are emerging and growing rapidly such as China or Brazil. New markets are opening to them, and the publishers have now opportunities to make more money in widening their market more than in intensifying the profitability of their present market.

NEW BUSINESS MODELS BASED ON E-ONLY

So, we would imagine new business models based on a new definition of holdings as electronic contents, and taking in account the role of the libraries in the preservation of these contents. These new models would be free from the reference to the costs of paper subscriptions.

Some studies are needed to set up these new models:

- on the cost of access to e-only. It is not related to the cost of a paper subscription, because access is not holding, and the production costs are not the same for e-only content and for paper content; only editorial costs are in common for the two types of materials.
- on the value of electronic content. The value of information is inversely proportional to its dissemination, and if value and cost were related, cost would decrease with the growing number of customers, of libraries paying for access.
- on the usage of e-journals by the researchers. A study for each title and of the titles more used and less used for each publisher, in relation with the cost of the subscriptions, would be useful, and would allow to approach some quality aspects in publishing.
- on the value of the subscription to a bundle of journals. The value of access to a bundle is far minor from the total cost of the titles of this bundle, as the usage statistics are showing that 80% of the downloads are made with 30% of the titles. In another way, as we cannot foresee to what titles the users would access, a cost for use model (pay per view) is not functioning and would be a barrier to a wide access to the information, and a subscription model to a bundle is more adapted.

- on the cost of archives. It would be minored to take in account the responsibility of libraries for long-term preservation and the value added by the libraries in preserving these contents. The preservation of electronic content is a major issue for our societies who need to guarantee a perennial access. This would recognize that cooperative actions in this field are supported by libraries and consortia, and that the long-term preservation is the job of the libraries.

- on regulations of multi-year contracts. These would take in account the evolving percentage of articles in open access in the content provided, the annual increases would be related to the annual inflation, reasonable (and not marginal) cancellations as in the "border warfare model" would be possible.

Consortia would cooperate in these studies, and publishers would also contribute to this action by making publicly available at the international level their business models, even if they take in account local criteria. A good new Big Deal would be one established by consortia and publishers in a more equal balance than the existing. If we don't progress that way, the public authorities will act and promote more competition in this market, and will decide more restrictive regulations in licensing tenders.

NOTES

[1] ArXiv provides in open access nearly 424.000 e-prints in physics, mathematics, computer science and quantitative biology.

[2] See: Press Release: New negotiation agent for NESLi2, 23 June 2003.

http://www.library.yale.edu/~llicense/ListArchives/0306/msg00099.html

[3] Cf the petition of European Union researchers that was signed at the beginning of the year by 25.000 individuals and institutions.

REFERENCES

1. European Commission: *Study on the economic and technical evolution of the scientific publications market in Europe - final report.* Brussels : EC Directorate C, January 2006, p. 5 - Comments from Elsevier. http://ec.europa.eu/research/science-society/pdf/scientific-publication-study_en.pdf

2. Gatten, J. and T. Sanville: "An Orderly Retreat from the Big Deal. Is It Possible for Consortia?", *D-Lib Magazine,* 10(2002)10. http://www. dlib.org/dlib/october04/gatten/10gatten.html.

3. Kiley, Robert: *Experience of Wellcome Trust in implementing an Open Access Policy*". Speech at the Conference: Scientific publishing in the European Research Area, Access, dissemination and preservation in the digital age. Slide 13. http://ec.europa.eu/research/science-society/document_library/pdf_06/kiley-022007_en.pdf

4. Kohl, David: *Ten years of Big Deals*, FECYT Seminar On Big Deals and Consortia, March 15, 2007, Madrid, Spain. http://www.heal-link.gr/SELL/2007/Madrid%20spr07a.ppt

5. SOFI e.V.- Soziologisches Forschungsinstitut Göttingen: *Survey on the Impact of VAT on Libraries and the Scientific Publication Markets - Final Report.* April – May 2006. http://www.sub.uni-goettingen.de/ frankfurtgroup/ vat/Endbericht VAT210906.pdf

7
CHAPTER

Developing European Library Services in Changing Times

Paul Ayris

Director of UCL Library Services, President of LIBER

ABSTRACT

The purpose of this article is to explain what academic and national libraries can do to continue to offer services and facilities at a time of economic difficulties. It identifies a number of methodologies and opportunities that are open to libraries and takes the view that it is never wise to waste a good crisis, because all threats are really opportunities in disguise.

The kernel of this paper was delivered at the 10th Anniversary special EISZ Consortium Members' Meeting on 2 December 2011, in Budapest, Hungary. It builds on an earlier talk delivered in Thessaloniki, Greece, on 14–15 November 2011 at the 20th Pan-Hellenic Academic Libraries Conference, entitled 'Academic Libraries and Financial Crisis: Strategies for Survival'. Both these presentations are available in UCL Discovery.

This article draws on themes used in both presentations, and introduces a new one on the topic of copyright reform. The article looks at the initial economic context for European research libraries and then examines ways in which libraries can tackle the threats which the current financial crisis poses. Joint procurement is one way in which libraries can achieve value for money, and the paper examines the role of JISC Collections in the UK. Innovation through collaboration and shared services are also ways in which libraries can innovate/make savings in a cost-effective way by sharing the burden of costs around the partnership. The paper gives two examples: one which is now well established, the DART-Europe portal for Open Access e-theses; and one which is in the early stages of being discussed — a cloud-based solution for true collaborative cataloguing amongst the UK's research and national libraries. The European Research Area (ERA) and the contributions that libraries can make to this infrastructure through innovative EU project funding are analysed in some detail by looking at LIBER's EU project portfolio. Finally, change and growth can come through changes to legal frameworks, and the paper looks at the Hargreaves review of copyright frameworks in the UK and the launch of the new library-based EU lobbying group for copyright reform, Information Sans Frontières (ISF).

Keywords: academic libraries; economic downturn; joint procurement; collaboration

THE CONTEXT

2011 has not been a happy time for European budgets and, at the time of writing, the Eurozone countries face fresh turmoil.[1] University funding is not immune from these changes in European society and, certainly in the UK, university libraries are experiencing cuts in recurrent funding. My own library has seen cuts of 13% over three budget years, with further annual efficiency gains expected. What avenues are open to libraries in situations like this to help them continue to offer services to support teaching, learning and research?

Joint Procurement

One avenue that is open to libraries is the possibility of undertaking joint procurement for content in a consortial setting — the reasoning being that there are savings to be made when a large number of players come to the table together to purchase content. In the UK, all Big Deals are negotiated for UK universities by JISC Collections. In 2010–11, JISC Collections realised £50 million in efficiency savings for British universities.[2] At a time of financial crisis, hard negotiations with all suppliers are expected. The current JISC model is an opt-in model, where JISC Collections negotiates on behalf of its members and then asks its members whether or not they wish to accept the deal. This is not the easiest way of negotiating, as the procurement team does not know (at the time of negotiation) how many libraries will sign up to a deal. This makes it a challenge to obtain the best value for money for libraries. However, the model does reflect the autonomy of UK universities in that they themselves make the final decisions over expenditure.

The current challenge with Big Deals is not the Big Deal model itself, but the amount actually charged for deals.[3] The benefits of the Big Deal are obvious:

- Desk top access to large amounts of scholarly material — and academics are certainly not going to complain about that opportunity

- Libraries welcome the model because of greater access and small libraries particularly benefit through access to a large number of titles which previously they could not afford

- Large publishers like it because it reduces attrition, resulting in additional e-fee income, locking libraries into multi-year agreements with built-in annual price increases.

In 2008, the JISC Journals Working Group commissioned Evidence Base to evaluate NESLi2 Big Deals in comparison with a title-by-title selection. They found that not only were the deals good value, in that previously unsubscribed titles were well used, but also that access to the Big Deal seemed actually to stimulate use.

However, there are disadvantages. In a Big Deal, libraries are unable to shape collections. There is less flexibility to respond to budget cuts: 'In Britain, 65% of the money spent on content in academic libraries goes on journals, up from a little more than half ten years ago.'Economist, May 26th, 2011.

Some content is not used and some journals in a deal are more valuable than others. There are difficulties in small journal publishers and monograph publishers being able to sustain sales in this environment, which in turn increases the market power of a small number of large publishers.

In a survey of a number of other business models Lorraine Estelle, the Chief Executive of JISC Collections, has concluded that no other business model currently challenges the predominance of the Big Deal. The high cost of journals is a consequence of continual increases in the number of articles published each year, the uncompetitive nature of the market place, and the result of market power. The academic community will face these issues irrespective of the business model.

Innovation through collaboration

Cutting costs through collaborative purchasing is one way to save money. Another way to support services is to innovate through collaborative partnerships, where the costs of development are shared around the membership. A good example of this sort of development is the DART-Europe portal, which gives access to European Open Access research theses. At the time of writing, the portal indexes 254,641 full-text research theses from 379 universities sourced from 20 European countries.[4] The portal is run by UCL (University College London) on behalf of LIBER members and the partnership commenced in 2005.

A new Report from the ERAC (European Research Area Committee) on National Open Access and Preservation Policies: Analysis of a Questionnaire to the European Research Area Committee appeared at the end of 2011.[5] It is a pan- European analysis of the state of play in Open Access and Digital Preservation, with the analysis being taken from answers to a questionnaire distributed in 2010 to all ERAC members and observers. The Commission received 29 responses between 21 December 2010 and 11 March 2011: 25 from EU Member States (Bulgaria and Hungary did not respond) and four from ERAC Observers (Iceland, Montenegro, Norway and Switzerland). In the area of Open Access, the Report notes that there has been significant work done on research theses. The DART-Europe portal is highly commended in the Report and is referenced explicitly in sections 1.4.2 and 1.5.5 as a model of good practice.

How was this work accomplished and what business model supports the development? The architecture which supports the service is portrayed in Figure 1. There are essentially three layers to the service — a storage layer (at the top of the Figure, usually Open Access repositories based in institutions), a harvester and a service which normalises the data (based in UCL), and the public portal (also based in UCL). Only metadata is harvested, not the full text of the thesis. The full text remains on the local server and the metadata points to it from the portal based in UCL.

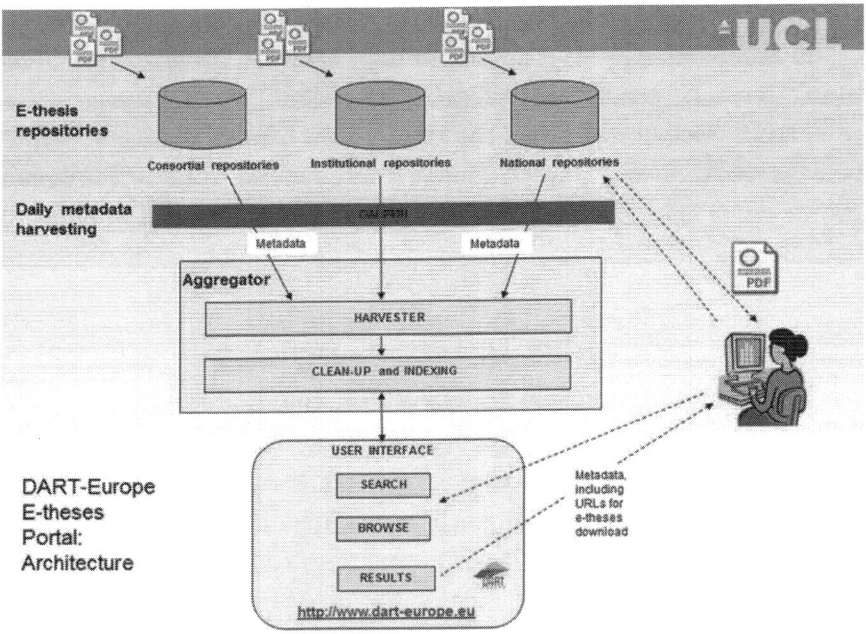

Figure 1. The DART-Europe architecture.

The partnership and business models which underpin this service are important and are a significant part of the reason why the models are successful. The partnership model is based on the 400+ members of LIBER (Association of European Research Libraries).[6] Partnership in the DART-Europe portal is based on the LIBER membership. There is no separate fee for contributing metadata to the DART-Europe portal. The ability to be a partner in the DART-Europe partnership for free is one of the benefits of LIBER membership. All partners are invited to sign a partnership agreement,[7] which outlines principles and values which the partners espouse. The first two principles cover how the partners contribute metadata to the portal:

- 'DART-Europe will encourage the creation, discovery and use of European e-theses, and will maintain a central Portal for e-thesis aggregation and access.
- 'European libraries and consortia are invited to contribute metadata to the DART-Europe Portal. Contributors will determine the terms and conditions under which their metadata are contributed.'

The business model is covered in the third of the principles in the partnership agreement:

- 'DART-Europe welcomes the contribution by partners of resources to support the management, discovery, usability and preservation of e-theses, and to further the aims and objectives of DART-Europe.'

What this means is that the costs of running and developing the portal are met by UCL, and the costs of contributing metadata to the portal are met by the individual partners, usually by work in kind in their technical departments. There is no cross-subsidy from LIBER for maintaining the portal — LIBER membership fees do not subsidise LIBER services, which are meant to be self-funding. Development work on the portal is currently funded by UCL as part of its own programme of IT development. It is an unusual business model, and it is not clear how scalable it is to other services. What underpins the partners' commitment to maintain and develop it, however, is the impact that access to European research theses has on those who use them.

Figure 2 shows the results of a comparative study in Dublin City University, a DART-Europe partner, of the use made of paper theses stored in their library and digital theses in the institutional repository.[8] The number of print loans includes all print theses stored in the library (almost 30 years). The number of e-thesis downloads accounts for only the theses that have been added to the institutional repository (less than 2 years). The figures speak for themselves. There were 518 print thesis consultations in 2009, compared to 16,212 electronic thesis visits. The conclusions these statistics suggest are supported by comments which are e-mailed to the DART-Europe e-mail address by users who have used the service. To pick just one at random: 'I really appreciate this project, it is very nice to be able to have access to so many documents.' — Sebastian, France.

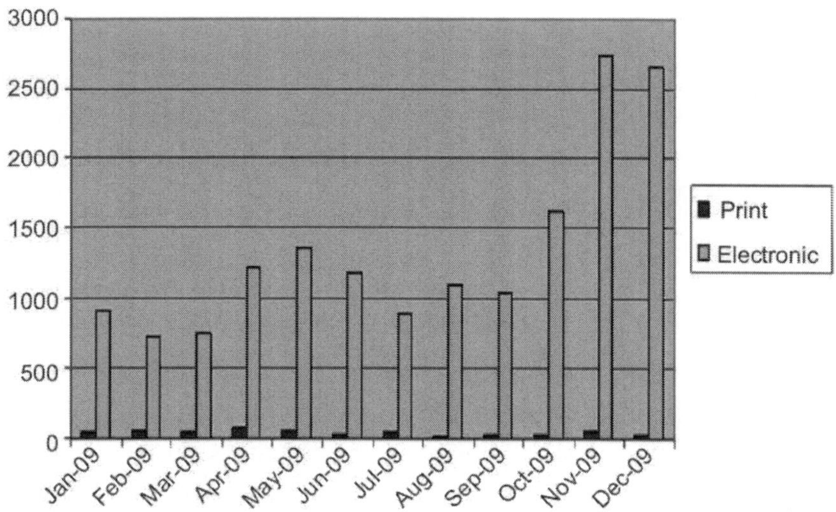

Figure 2. Use of print and digital theses, Dublin City University, 2009.

	September 2011	September 2010	+ / -
Records	220,203	176,189	+44,014
Universities	356	296	+60
Contributors	78	53	+25
Countries	20	18	+2
Languages	31	30	+1

Figure 3. Table showing the growth of the DART-Europe portal and the 20 contributing countries.

Rank	Country	Visits	Rank	Country	Visits
1	Germany	30,912	11	Iran	4,789
2	Greece	19,456	12	Portugal	4,280
3	Spain	15,442	13	Hungary	3,648
4	France	15,079	14	Algeria	3,187
5	UK	13,735	15	Serbia	3,162
6	Poland	13,228	16	Canada	2,999
7	United States	11,076	17	Sweden	2,846
8	Lithuania	10,059	18	Finland	2,708
9	Italy	6,895	19	Belgium	2,637
10	India	5,383	20	Turkey	2,523

Figure 4. Source of calls to the DART-Europe portal September 2010–September 2011.

The most popular countries from which the calls to the DART-Europe portal were made were:

The success of the DART-Europe portal shows that it is possible to innovate through collaborative partnerships to add value to the end-user experience. The keys to its success are the partnership model, which underpins the consortium, and the shared values in wanting research theses from Europe's universities to gain the maximum possible visibility in a networked environment.

Shared Cataloguing: the Next Generation

2011 saw much activity in RLUK, the UK consortium known formally as Research Libraries UK.[9] The 2009 Research Information Network (RIN) Report Creating Catalogues: Bibliographic Records in a Networked World[10] found that the current arrangements for producing and distributing bibliographic data for both books and journals involve duplications of efforts, gaps in the available data, and missed opportunities: '...[T]here would be considerable benefits if libraries, and other organisations in the supply chain, were to operate more at the network level.'

The RLUK Report, presented at the RLUK Members Meeting in London in December 2011, and currently being further studied by the RLUK Board, used the RIN conclusion as the starting point for a broad overview of current cataloguing activities and services in RLUK libraries. The Report, for which I had the honour to chair the Working Party, recommends that RLUK metadata is released as Open Data and that RLUK develop a more rigorous position on Linked Data, which would allow structured data to be linked on the web.

Were such changes to take place, the library catalogue would become re-positioned in terms of its relationship to the wider context of the web, and the social network of links that the web represents. The drivers for considering such developments are potential cost savings by RLUK members and a better service for users.

The Report identified that the best solution was for a cloud-based implementation to stand in for both local and central management of systems. This would allow local library management functions and a centrally shared metadata catalogue (community zone).

Metadata issues will need to be addressed: duplication of records for the same item needs to be replaced by the concept of a Master record.

The RLUK databases (COPAC,[11] for public access and the record retrieval service for copy cataloguing) need to be re-positioned in the wider context of the web. Coverage certainly needs to be expanded to include new media types, e.g., blogs, wikis, Open Access content, and e-books. A truly shared cataloguing service would reduce the footprint of local library management systems and it would re-define how libraries work.

The Report made a number of top-level recommendations for RLUK to consider further:

- that funding is identified to investigate the requirements and feasibility of a shared UK cataloguing service
- to co-sponsor with the JISC (Joint Information Systems Committee) a full cost-benefit analysis of providing an overall, above-campus, shared cataloguing system solution.

In the light of the 2009 RIN Report, and of the potential for network developments to transform the way that libraries work and the services they offer, the Report offers an exciting and challenging way forward for libraries to re-position themselves in the information landscape, with possible budget savings as a result.

European Research Area — Libraries as Part of the European Research Infrastructure

A particular area of work for LIBER is raising awareness amongst EU institutions of the potential for Europe's research libraries to play a bigger role in Europe's research infrastructure, and to be funded for this role.

At the LIBER Annual Conference in Barcelona in 2011, Commissioner Neelie Kroes acknowledged the role that LIBER is playing in helping to create a new information research infrastructure in Europe. She also set the ambitious target that the whole of Europe's cultural heritage should be available in digital form by 2026.[12]

One way that libraries can gain new income is by bidding for European projects. Unlike other funding agencies, the European Commission is significantly ramping up funding for research and innovation over the next 10 years — the proposed budget for Horizon 2020 (successor to the current funding programme, FP7) is €86 billion, an increase of 46% compared to the current programme.[13]

This is a lesson that LIBER is learning. LIBER has 420 members from across Europe and is a source for network building and collaborative partnerships. LIBER now has a full-time EU Projects Officer and, from January 2012, an EU Projects Communication Officer. The current LIBER EU projects portfolio has grown from zero in the last three years. European projects in which LIBER is/has been involved, are:

- Europeana Travel — a digitisation project, which finished in May 2011, with 1,000,000 digital objects on the themes of travel, tourism and exploration being made available to the Europeana portal.[14]
- Europeana Libraries — LIBER's biggest EU-funded project to date at €4,000,000. Output will be a new aggregator for metadata and full text from Europe's research libraries into Europeana. 5,000,000

new digital objects from research libraries will be made available in Europeana by 2012.[15]

- APARSEN — looking at the start of preparations by stakeholders across Europe for digital curation. Builds on LIBER's input into the US Blue Ribbon Taskforce on Economically-Sustainable Digital Curation.[16]

- ODE — Opportunities for Data Exchange. Looking at the level of preparation in Europe for research data curation and re-use. LIBER will manage the input of European research libraries.[17]

- Newspapers Online — a project which starts in January 2012 with the aim of adding 29 million pages of European newspaper content to Europeana.

- MedOANet — starting in December 2011 with the aim of facilitating Open Access policies and strategies in Mediterranean and neighbouring countries; kick-off meeting in Athens in January 2012.

- AAA — a technical project looking at (amongst other things) the opening up of computer networks across European borders which will allow researchers away from their home university access to their subscribed digital content.

For all LIBER's confirmed projects, and projects under consideration, the income for the LIBER Foundation alone is €521,597. This is new income, which would not otherwise be available to LIBER. The purpose behind LIBER's bidding for such funding is to embed research libraries into the European research infrastructure. An issue is always the percentage of total costs which the Commission will pay. LIBER will only bid for projects which are 75%–100% funded. All LIBER members can become partners in LIBER projects, and the LIBER membership fee is a maximum of €425 per year.[18]

Pushing Forward the Borders

LERU is the League of European Research Universities,[19] an association of 21 leading research-intensive universities that share the values of high- quality teaching within an environment of internationally competitive research. Founded in 2002, LERU is committed to:

- education through an awareness of the frontiers of human understanding
- the creation of new knowledge through basic research, which is the ultimate source of innovation in society
- the promotion of research across a broad front, which creates a unique capacity to re-configure activities in response to new opportunities and problems

The purpose of the League is to advocate these values, to influence policy in Europe and to develop best practice through mutual exchange of experience. LERU regularly publishes a variety of papers and reports which make high-level policy statements, provide in-depth analyses and make concrete recommendations for policymakers, universities, researchers and other stakeholders.

One of the areas in which LERU has begun to work, is Open Access. A general meeting of LERU Chief Information Officers/University Librarians took place in UCL in London in December 2009. That meeting appointed a Working Group to draw up a LERU Roadmap towards Open Access. The Roadmap was considered twice by LERU Vice-Chancellors at meetings in London and Paris and it was launched in Brussels on 17 June 2011.

The purpose of the Roadmap is to offer guidance on how to position a university in the European Open Access landscape. It builds on the

Open Access Statement of the European Universities Association[20] and is a Roadmap for all European Universities, not just LERU members. The Roadmap[21] offers advice on the following issues:

- Open Access in a wider context: Open Scholarship and Open Knowledge
- The Green route for Open Access — Steps to Take
- LERU and the Gold route for Open Access
- Models of Best Practice to support the Roadmap
- Benefits for researchers, Universities and Society

The two basic mechanisms through which researchers can make their work freely available are often termed as the 'gold route' and the 'green route'. The adoption of either or both routes could lead to a transformation in the means of disseminating research outputs by LERU and other universities across the globe.

LERU and/or other universities can consider having Open Access repositories into which, copyright permissions allowing, copies of their members' research outputs could be deposited. Those who already have such repositories are continuing to develop them. Many universities have found the Green route a helpful one to follow as a means of improving the dissemination of research outputs. In Webometrics listings of the impact of institutional repositories, LERU universities are significant contributors. The July 2010 listing showed that five of the top ten European universities listed are members of LERU.[22]

Several universities have supported the Gold route for Open Access, whereby authors in these institutions either publish in Open Access journals or pay publication charges (funded by the research funder or from an institutional Open Access fund) to make their article available in Open Access on publication. Some research funders, such as the Wellcome Trust in the UK, the Austrian Science Fund (FWF) and the Netherlands Organisation for Scientific Research (NWO), will fund such publication payments. The Gold route is a bold route, which may also change the pattern of publication.

In the wake of publishing the Roadmap, some LERU members have become more interested in Gold Open Access publishing. A meeting of interested parties took place in April 2011 in Oxford where a wide range of issues was discussed. A follow-up meeting of LERU Chief Information Officers took place in UCL in London in November 2011. One of the action lines which the meeting agreed was to form a LERU Working Group to look in greater detail at Gold Open Access. In particular, the meeting was interested in investigating further the possibilities for shared Gold Open Access infrastructure across LERU Universities.

The Chief Information Officers noted that the OAPEN project [23] has as its objective to develop and implement a sustainable Open Access publication model for academic books in the Humanities and Social Sciences. The OAPEN Library aims to improve the visibility and usability of high quality academic research by aggregating peer reviewed Open Access publications from across Europe. The meeting of November 2011 also discussed in detail a library consortium business model from Dr. Frances Pinter, which aims to create a new business model for Arts and Humanities monograph publishing. [24] The LERU Working Group will also consider this Business Model in more detail to see whether they can adopt such a model to further the Gold Open Access agenda.

COPYRIGHT REFORM

In a digital environment, one of the frameworks in which libraries, researchers and teachers have to work is copyright. Copyright legislation, at the Member State or at the EU level, is the framework in which people act. In the UK, the new Coalition Government sponsored a review of the UK's copyright framework by Professor Ian Hargreaves.[25] One of the drivers for the review was to see whether the UK's Intellectual Property framework needed to change in the interests of innovation and growth. The answer proposed by the review is that it does.

In relation to personal copying, the Report has this to say:

1. 'Copying should be lawful where it is for private purposes, or does not damage the underlying aims of copyright...

2. 'The UK has chosen not to exercise all of its rights under EU law to permit individuals to shift the format of a piece of music or video for personal use and to make use of copyright material in parody. Nor does the UK allow its great libraries to archive all digital copyright material, with the result that much of it is rotting away. Taking advantage of these EU sanctioned exceptions will bring important cultural as well as economic benefits to the UK. Together, they will help to make copyright law better understood and more acceptable to the public. In addition, there should be a change in rules to enable scientific and other researchers to use modern text and data mining techniques, which copyright prohibits.'

The review had similar strong things to say about Orphan Works.

'It [the UK Government] should begin by legislating to release for use the vast treasure trove of copyright works which are effectively unavailable — "orphan works" — to which access is in practice barred because the copyright holder cannot be traced. This is a move with no economic downside.'

The Report suggests that, if all its recommendations are adopted, the result will be stronger rates of innovation and increased economic growth. An economic impact assessment conducted by the review team, and of course subject to the high degree of uncertainty inherent in such projections, estimates that this would add between 0.3 per cent and 0.6 per cent to annual GDP (Gross Domestic Product) growth. The path laid down in this review would also, over time, mean that intellectual property law, including copyright law, would become clearer and be observed by most people without controversy.[26]

Spurred on by developments such as the Hargreaves review, European research and national libraries have begun to lobby for copyright reform across Europe. 2011 saw the creation, with private sponsorship, of Information Sans Frontières (ISF).[27] Amongst its members are the JISC (Joint Information Systems Committee) in the UK, LIBER, Europeana, and EBLIDA. IFLA is an observer on the Steering Committee. ISF was formally constituted in September 2011. It has two advocacy officers, based in the UK and Brussels, who work with the EU institutions to put the case for academic-friendly copyright reforms across Europe. The first work of the ISF is on the European Commission's Proposal for a Directive on certain permitted uses of orphan works with a view to establishing common rules on the digitisation and online display of so-called orphan works. The ISF has laid down reasons why it believes that the draft Directive should be re-cast. Time will tell whether they are successful.

CONCLUSIONS

What conclusions can be drawn from this overview of European library activity in a time of economic crisis? Joint procurement can deliver savings through bringing a large number of libraries to the negotiating table together. In this way, it is possible to do more with the level of resource that is available. Innovation through collaboration and shared services is certainly a fruitful area for future growth. The DART-Europe portal provides access to 250,000 research theses and is a successful collaboration amongst several hundred European universities. RLUK is beginning to look at new models of deep sharing, based in the cloud, for collaborative cataloguing which could deliver savings to members and avoid much of the waste and duplication that is inherent in the present system. There are new sources of income available to European research libraries, but for many libraries this is a new development and would need to become embedded into their organisational structures. Finally, the paper shows that growth and development can come from changes to legal frameworks, and looks to the Hargreaves review in the UK and the new work of Information Sans Frontières (ISF) for indications of what is possible.

NOTES

1 See http://www.guardian.co.uk/world/2011/dec/26/euro-crisis-2012-

eurozone?INTCMP=SRCH.

2 See http://www.jisc-collections.ac.uk/Documents/JISC%20 Collections%20 Annual%20Review%20Full%20Accounts% 20 version %202011.pdf.

[3] For this and what follows, see L. Estelle, 'Big Deals and Alternative Business

Models', delivered at the 2011 annual meeting of the Chief Information Officers in

LERU Universities (League of European Research Universities) at http://www.leru. org/index.php/public/extra/cio-28112011/.

[4]http://www.dart-europe.eu/browse-list.php?index=institution, http://www.darteurope.eu/browse-list.php?index=country.

[5] The Report can be found at http://ec.europa.eu/research/science-society/

document_library/pdf_06/open-access-report-2011_en.pdf.

[6] See http://www.libereurope.eu.

[7] See http://www.dart-europe.eu/About/info.php.

[8] See Hill, R. and Moyle, M. (2010), 'Investigating the Impact of e-Theses at DCU',

Presentation at the 2010 LIBER Annual General Conference, http://discovery.ucl. ac.uk/19955.

[9] See http://www.rluk.ac.uk.

[10] See http://www.rin.ac.uk/our-work/using-and-accessing-information-resources/

creating-catalogues-bibliographic-records-network.

[11] See http://www.copac.ac.uk.

[12] Available on YouTube at www.youtube.com/watch?v=GIU14-3hYto.

[13] See http://www.ucl.ac.uk/news/staff/staff-news/05122011-mbrowne?dm_ i=UAA,NDYU,3YRLM9,1W32P,1, where Michael Browne (Head of UCL's European Research and Development Office) analyses the importance of European funding for universities.

[14] See http://europeanatravel.eu/.

[15] See http://www.europeana-libraries.eu/.

[16] See http://brtf.sdsc.edu/.

[17] See http://www.d-nb.de/eng/wir/projekte/ode.htm.

[18] For further information on LIBER's EU project portfolio, see P. Ayris, 'Establishing the Library Landscape in Europe: LIBER's Portfolio of EU projects' in LIBER Quarterly, vol. 21 2011 (1).

[19] See http://www.leru.org.

[20] See http://www.eua.be/eua-work-and-policy-area/research-and-innovation/ Open-Access.aspx.

[21] The Roadmap is available at http://www.leru.org/files/publications/LERU_AP8_Open_Access.pdf.

[22] See http://www.webometrics.info/top100_continent.asp?cont=europe.

[23] See http://www.oapen.org/home.

[24] See http://www.youtube.com/watch?v=niyYWVa2w6w.

[25] See http://www.ipo.gov.uk/ipreview-finalreport.pdf.

[26] All the quotations in this section are taken from the Executive Summary, pp. 3–9.

[27] See http://informationsansfrontieres.eu/.

8

CHAPTER

The Multimedia Library as a General Perspective of the Library System

Lev Amlinski

Scientific & technical information centre – InformA Consulting of the interior lay-out of libraries An der Kolonnade 8 10117 Berlin, Germany

Globalisation of the information transmitting system in its complexity of scientific, educational, social and economic demands of society has lead to the formation of the international information market. Libraries have always held an important position in the entire information market and will continue to do so in the future. When talking about the development of libraries it is important to come to terms with the most important scientific-technological and economic realities of the present time:

- the constant appearance of new information carriers on the market;
- the fast development of hardware and software;
- the constant development of new communication technology;

- the increasing competition on the market of information performance;
- the increasing educational factors on the job market.

The present situation of the modern library system differs from all previous situations: the library technologies are undergoing active digitalization.

Multimedia has developed as a reality, that at the same time has an effect on rational thinking and on emotional creation. This will eventually lead to a qualitatively new perception of events and appearances of our surroundings and will activate associative thinking amongst other things.

The new thing in principle of multimedia formation is that - instead of the former well-known technologies - now the numerical technology on a computer basis is used for the co-operation with different analogue media (cinema, TV, etc.). Thus we will gain a synthesis of information of all carriers and transmitters.

The new thing in the utilisation of multimedia is not only the possibility of receiving messages from different media but also the possibility of interchanging during correspondence in an optimal way for every adequate variable level. In this way the multimedia systems will develop into a multichanneled communication system.

the fact that people know the possibility of choice of the needed fragments of every information media and after that can again synthesize, complete, change and add new elements and thus create their own multimedia interface fitting their own wishes and demands. In this way it is possible to develop individual educational, scientific, entertaining and other groups of information on the basis of an electronic book for example or any other apparatus.

Nowadays and in the near future, libraries will give consideration to multimedia.

Having dedicated my presentation to the topic of multimedia libraries I will now go on to explain this term from my point of view.

Although this term is now often found in specialist literature, nobody really can say how this library has to look in order for it to be considered a multimedia library. We can theorize a lot, and discuss the terminology, but in my opinion the most important questions are: What do these libraries do for the readers and how do they effect the job of a librarian? Concerning the cooperation between the librarians and the architects one has to realize what kind of instructions the librarian gives to the architect when planning a multimedia library. It is also important what kind of specification the inner architecture of the library building should have.

A multimedia library is one that has a multimedia stock, technical equipment and computer nets, that can be used by the reader. Moreover they provide the reader with the possibility of entering the net and a complex perception, both acoustically and visually, of the information space, the search for the required information in this space and working with this information.

One has to confirm that even the term multimedia differs from its common etymological origin, that is, multimedia is more than the sum of its parts. This is because multimedia also has the computer numerical display at its disposal which in principal provides new quality and - at the same time – provides for the librarian's demands, i.e. using the multimedia stock. The libraries possess different media in analogue display, these monomedia can be revised by the libraries and worked out as multimedia for the readers.

Here we have to point out that - by no means – will the libraries give up their traditional technologies, their traditional stock and the traditional condition of the readers and they don't part from their vast experience either.

One important claim for the libraries is their providing a psychological comfort to the reader getting deeply involved in multimedia and further they consider the specifics of behaviour and of the psychological situation occurring under the concrete condition of the readers using the library with all their different individual interests.

Now let's explain the main categories of the above definition.

MULTIMEDIA STOCK AND MULTIMEDIA CATALOGUE

The strongly increasing interest in multimedia in the fields of science, education, culture and entertainment is due to man's unlimited possibility of taking part in the formation, changing and utilisation of multimedia material - including elements of playing games. Just think about computer games having developed to a vast economic branch. As Beate Tröge says in her interesting article (B.I.T. 2 (1999) Nr. 4), during education a person remembers until the next day - only 30 % of the information that he has read, 20 % of the information he heard and 50 % of the information he heard and read at the same time. However - he or she remembers 90 % of the information which he has - at the same time - heard, read and revised by himself, changed and added own ideas to. Concerning the production of multimedia information and their application in high school learning - the author maintains - libraries should play the biggest part in this and should co-ordinate with information centres and computer centres.

What about the opinion of the libraries of this statement?

During an international conference about the media-competent library in the information society "Die medienkompetente Bibliothek in der Informationsgesellschaft" (6. - 10. Dez. 1998, Berlin), as a result of interesting and multiple discussions it turned out that a librarian always has to attend further training to be able to master the whole multimedia technology. Also he should learn how to be helpful to the reader researching the required information. Libraries themselves should develop into an "information store" and the whole tendency of the development of the librarian should run in the direction of an "information broker". All this is very interesting, but we shouldn't forget, that we only can get to know what exists and that it is absolutely impossible to get to know what doesn't exist at all. In the most libraries there is hardly any multimedia technology up to now. This technology has to be developed. This technology cannot be lent - e. g. from the mass media - because the vast TVbroadcasters are interested in having the numbers of professionals limited. Contrary to this in a library this technology has to be used by thousands of people of different educational levels. The readers are interested in all fields of social knowledge. Thus we need a specific library multimedia technology.

The librarian can take part in this process as an expert, and also as a user, but not as a developer of this technology. If we try to change the librarian into a programmer, into an art director, into a sound engineer and last but not least into an expert of art , we never will reach this aim but we will lose him as a librarian. This is why - if the library wants to build up a real multimedia stock and wants to offer it to its readers – we should first of all develop a new technology. This technology should be professionally dealt with by several specialists.

It may be that some library directors will oppose this thesis by arguing that the development of experts is principally not possible, that it cannot be afforded because of the lack of financial means due to the current severe economic situation.

But if we want to have multimedia libraries, it is absolutely essential to build up such groups as groups of programmers and computer experts which were hired 20 -25 years ago.

Concerning the libraries, they are very successful, master the new work with the information carriers, with the computer net and the rhythm of a librarian's job getting faster and faster.

Everybody knows that every library starts with its stock. Exactly the same applies to the multimedia library. Multimedia materials that are used in the library can be split into three parts:

1. these are the products which come complete from different producers such as publishing houses, TV-studios, cinema-studios, multimedia centres etc.;

2. products that are produced in form of multimedia blocks according to individual orders or demands;

3. products that are produced by the readers themselves, either individually or in groups including teams communicating via the internet.

The first above-mentioned group has no individual address. Whereas both of the next groups always have an individual address and it is especially this individual address which causes the complicated character of the libraries' multimedia technologies.

Of course it is possible to undertake another classification of the stock but besides this we must have the necessary digitized catalogues along with the multimedia stock. That is, when talking about multimedia stock we will automatically come to the multimedia catalogue.

The multimedia catalogue should consists of four parts:

- hypertext catalogue;
- audio catalogue;
- catalogue of fine and plastic art;
- catalogue of video films.

The hypertext catalogue contains bibliographical data of books, journals and various other texts, also the written data of audio, video, fine and plastic arts and others. The hypertext catalogue disposes of a wide-channeled structure of bibliographic data with many crossovers. But we also have to consider that we cannot integrate a piece of music, an art work, a video film into a multimedia block if we only have the written data of these works. The music has to be heard, the art works and the videos have to be seen, which means that the multimedia catalogue should contain music, an art work or a film. All this has to be digitized.

These parts of multimedia catalogues can be linked with the hypertext by special options, but nevertheless they keep independent of them at the same time. The complete structuring of these additional catalogues like e.g. music catalogue has to be worked out according to its attributes (e.g. stilistic attributes like classical music, popular music, light music, pop, techno, jazz; geographical music like European music, Latin American music, African music etc.; instrumental attributes like piano, violin, orchestra and chorus). Of course it is possible to detail all characteristics even more precisely to make this classification more convenient for the user.

Improving the convenience of the user is one of the most important requirements of a multimedia catalogue - in my point of view. Concerning this we can state that the digitized multimedia catalogues of the library are not very convenient to users. Of course they offer different ways of approaching: via author, key-words, subjects etc., but besides this useful efficiency the user cannot make any additional precision work nor assorting of the chosen documents within the vast list of found documents. The further development of electronic catalogues like the multimedia catalogue is important because during the last decade especially the catalogues were the quintessence of the complete computer technology in the libraries and still they are. Actually electronic catalogues are a shop-window for computer technology and they connect the reader with the interior technology of libraries. If we look at this from the angle of an eye of the user's convenience we should consider the multimedia catalogue as the highest level on the fields of electronic catalogues, because this catalogue can deeply intervene in the creative process whilst working on multimedia products.

Since four to five years active research on the fields of metadata are conducted. Metadata differ from the traditional catalogue data by the possibility of delivering the document with the support of the proper software, which shows the address of the document in the net.1 Certainly this is not the only difference but this links the document with the electronic net, in which it is. Thinking of a future fusion of multimedia catalogue with metadata we can explicitly refer to Amanda Xu whose statement of opinion is, that electronic catalogues of libraries can function as a sluice to the electronic resources in Internet.2 If libraries begin to use this sluice, they will gain a real possibility to improve their performance for the readers multifariously.

The multimedia catalogues can make the development of multimedia blocks easier and thus they push forward and support the development of combination theory within a creative process. At the same time multimedia catalogues are a real technological base for generating the multimedia products in libraries.

THE INFLUENCE OF MULTIMEDIA ON THE INTERIOR DESIGN OF LIBRARIES

There are disagreements between the prediction of scientific-technical development and the development in itself. That is because the predictions are based upon the existing scientific-technical level, upon the categories that are already known. But the development itself doesn't run linear, but irregularly as a result of new knowledge, principally new products etc. All this applies to its whole extent to the libraries. And looking from this point, what has been performed by the librarians and architects during the previous decades, we can state that especially the extreme uncertainty about the libraries' development had an effect on the libraries' inner decoration as the most important point.

That led to its concrete solution in projecting space of extreme flexibility. That has saved libraries from rapid becoming out of date and this will remain an important solution in this sense for the future, too. But it is already clear now, that the high flexibility of the space is not enough.

What kind of flexibility can be helpful when at the same time the stock, the number of readers and the number of employees is increasing? In the years between 1960 and 1993 at the university library of Münster, the personnel has more than doubled, the stock almost tripled, the number of readers has increased seven times and the loan of books has increased nine times. But at the same time there are several libraries that don't change so extremely, that cannot even say how the numbers of readers, the stock, etc. alter.

The question is why there is a necessity for new principles of functional architecture when building libraries. To answer this question - in my view - we should point out, that all previous changes of functional architecture, including the fixed planning in three sections and their later flexibilisation, are connected with the printed paper as information carriers like books, journals, newspapers, catalogues, loan forms etc. At the time of the introduction of the fixed planning of three sections of libraries there was neither mechanization nor automatic control, at the beginning of the flexible planning of libraries there were no personal computers or digital information networks. Thus the transmitting of information was bound to the physical circulation of the information carrier. The choice of books and their handing over claimed the movement of the reader in the reading zone. The whole job of the library's staff was and still is practically connected with its movement between all library zones. That leads to our statement that the inner flexibilisation of every library zone was and still is the most effective solution to stop the becoming out of date becoming out of date process of libraries.

But at the same time we have to realize that the new realities in the information society, including new means of mechanization and automatic control, new means of transmitting, new information transmitters and -nets will bring up suitable principles for the library architecture. When Gerhard Liebers said twelve years ago, that the future development of libraries cannot be predicted, then now such a statement has become even more relevant with the increasing dynamic of the whole development of the information society. In this context one should do without the complex planning of the building and without short ways of transporting books and slips. A functional inner architecture of library buildings will be of utmost importance: the planning of large-scale spatially-independent zones of the library buildings which open an independent possibility of development suitable to the changing of their functions.

The planning of the large-scale spatially-independent zones of the library buildings as an architectural principle makes it possible to maintain good performance of libraries for all future variations of the libraries' position on the information market under the condition of rapid change of its sociocultural functions in society. In the long run it will be up to the libraries to escape from the sword of Damocles of the choice of alternatives, of being forced to choose between the short-termed moral age of library buildings or the frozen financial budget for the prolonged maintenance of their existing building structure.

In general we presume that when it comes to planning, it is especially the librarians who support the ideas and solutions of functionality of inner decoration.

That leads to the question, whether library functions are considered only to be technological processes and operations that immediately focus on the condition of the readers, or whether a wide range of tasks belong to the functions of a library, including the supply of the correct psychological atmosphere for visiting a library and working inside, becoming aware of taking part in the reality of the building and appreciating the library as a temple of science. In between the wide range of opinions about the role of the university libraries, for example, within the complete structure of the educational and learning process as within the scientific work at universities there are diametrally contradicting views. One position has the opinion that libraries have to be more integrated in the university campus and so will become a kind of centre for meetings, communication and discussion. The other position maintains that – first of all – a library should support the meeting with the book, should support systematically working with the aim of gaining knowledge and that it is not a club for any meetings equipped with the proper interior decoration of the reading zone. Assuming that as a rule the truth is in the middle, we notice that the task of the libraries is to attend a vast range of interests of the readers. By realizing this enormous part of libraries the whole space of the reading zone plays an important role, also the height of the readers' equipment, their fresh air and light supply. Thus the best libraries became public treasuries of world's architecture.

At present there remains the challenge to find out the best possible partition of the readers' zone. This has to be done under the aspect of the current tasks of library service, of the increasing numbers of readers, of their changed demands under the conditions of an information society and also under the aspect of the further development of the library's service as an organic part of a homogenously structured information market and information service.

Also it is clear, that the reading zone in the digitized libraries and especially in multimedia libraries is getting larger because the technical equipment of the readers' place is increasing. And if we now have only laptops and computers in the reading zone, we have to be aware that a place equipped with multimedia needs a lot more space. This doesn't relate only to the technical provision but also to the privacy of the reader, his or her productive work in a multimedia library. This privacy has to be protected by using one or more screens, either by screening or by a special lamination of the screen or by special glass. Between 1960 and 1969 a famous American psychologist, Robert Sommer, undertook research into the psychological condition of the readers in the reading-room. In this study they found out, that every reader who comes during the first hour of opening chooses a place without a neighbour. Only later on when a lot of places are occupied, does he have to look for any place left. Everybody dealing with multimedia in a private creative process always prefers privacy, nevertheless on the contrary it is not proved that single cabins are more advantageous.

The complicated character in the present phase of the libraries' development is the fact that the information society, which we live in, has produced new markets, new products, new ways of communication. If we want to characterize the current situation of this information society with one single word, it would be – globalisation. Globalisation of financial markets, globalisation of communicative connections via the internet, globalisation of economics, globalisation of the job market. But if we project the complete globalisation on man, then the effects of it will turn out as a globalisation of his information survey. And, above all, it will make life more convenient. Globalisation can be only advantageous for people if it brings with it more comfort. This comfort is not a state of rest. This comfort can be considered as being in accordance with the outer living circumstances and the inner need.

But now let us talk about the readers. What does a convenient situation mean to the readers? What does the readers' comfort mean to the libraries?

A market of information has come up. Libraries have a definite position in this market. They shouldn't lose this position. If libraries want to remain competitive in this job market they have to provide a certain level of convenience for the readers. The comfort for the readers is not a generosity of the libraries but it is an attempt to get a chance for survival.

All this shows us that we have to thoroughly consider the whole equipment of the reading zone. Today it is no longer a rectangular room with tables and chairs in rank and file to have room for a maximum number of readers. In cooperation with librarians, architects and psychologists the reading zone can be divided in three areas, but without fixed borders:

- the zone of the intense psychological co-operation of the reader with other readers;
- the zone of partly enclosing in to this mentioned above psychological intensity;
- the zone of complete psychological private sphere (psychological demarcation).

All three of them of course have to be variable. But all three of them should be freely available to the reader.

Precisely this principle of the reading zone for the "individual reader" instead of the reading zone for the "average reader" should - in my opinion - be the decisive principle while planning a reading zone. In the reading zones some glazed cabins should also be installed for multimedia info-studies in which a team of different professionals together can produce a multimedia block with a computer network.

We should never try to put the reader under pressure.

When - for example - the architect Kahn whilst projecting the University Library of Exeter (New Hampshire) planned niches in the reading zone for the individual use and for intensifying the reading work, the students considered it to be a refuge for hard work and refused them in their original intention because they needed communication and eye contact.

Kahn came to the conclusion – according to Glenn Robert Lym - that the chronic locality of a library is determined by the individual being inspired by a book. The book is taken out of the dark archives into the sun light. Thus the sun light also has to enter the building. Then the reader enters the niche, which is the beginning of the architectural structure that the user becomes aware of. The reading zone is impersonal. Although the books are in the dark, they should be visible and available for stimulation. That can be supported by the open shelves inviting use.

Nevertheless a proving enquiry of users showed that the students didn't accept the architectural structure of Kahn (sunlight and niches). Kahn assumes that the reason was, that the students didn't consider the libraries' books in the same way as Kahn did, not as a ritual of inspiration but as a very dry job that has to be done quickly. In this sense the students think of the niches as places of hard work and studies. Concerning the presentation of books - very few of them considered them being inspiring, others even found them off-putting

The students of Exeter Library don't look at the library as a place of lonely inspiration by and with the book, but as a place of communication. The social centre of the "campus" has moved to the library.

According to this experience - in the opinion of Glenn Robert Lym - the library should have: a department of reference, a department of books and reading and kind of a social department for meeting people without disturbing the others.

That is, the central department of a library should not invite to read via books, although turning over the leaves of free available books is a good thing and should be maintained in the future, but the library should additionally be used as a meeting point.

Different readers need different conditions ranging from solitude to high frequency in the reading zone. Also one and the same reader needs either this or that at different times. I am not sure whether the principle of projecting the reading zone for the average reader is the optimal solution.

Part of multimedia are the electronic books, which in principle supply new possibilities for the reader. Clemens Deider and Rolf Fuhlrott worked out a detailed statement about electronic books.3 All electronic books have advantages but only one is of revolutionary importance, namely the one developed by assistant professor Joseph Jacobson at Massachusetts Institute of Technology (MIT). The pages themselves were written upon with electronic ink. This e-ink is surrounded by a special foil of 0.2 millimetres thickness and consists out of an innumerous amount of capsules about 0.04 millimetres in diameter, that move between two transparent electrodes. By making them alive, black and white particles of about 0.1 microns move inside the micro capsules to the opposite side, meaning that, according to the order, the capsules turn round to the black or the white side and thus form the desired letters. According to the developer - the MIT assistant professor Jacobson, a resolution of 1,200 points pro inch can be reached with the black and white version. Up to now 150 dpi has been enough for him. As there is no need to permanently build up the pages with the electronic ink, the energy required is minimal. Even without electricity, which has to flow when writing or renewing pages, everything can be read black on white. It hasn't yet been determined how long the face remains stable, and doesn't fade, thus needing to be refreshed. Besides this success the scientific searchers at MIT are already working on a color electronic ink.

According to Joseph Jacobson, up to 500 megabytes can already be saved. While using different flat memories, it would be possible in future to accumulate up to 35 - 350 gigabytes after a developing proper flat memory card with a quantum-mechanic effect. Talk has started about the future possibility of saving the complete Congress Library onto an electronic book. It doesn't look so bad for advertising, but the important thing for us is, that also static and moving pictures and also music will be accumulated in this electronic book at some point. Readers will get a multimedia carrier which they can use in different ways.

Nowadays there is a discussion about the fields where electronic books can be used. I think this discussion is a bit premature. Without referring to the different positions of the discussion, the electronic books will make their own way as to where they will be successful. The most important thing is that these electronic books as multimedia carriers will be able to change the complete educational process, the whole didactic. These books can bring democracy to the educational system because the scholars and students will be able to take an active part in structuring the teaching material. Everybody can find out their suitable learning strategy. They won't have to follow a certain teacher or a certain professor all the time.

Of course that means, that libraries have to take care about the electronic books, these multimedia books being structured according to the sense of their content. This is not really a new function of libraries, because libraries have always acted as the bibliographic reference for the readers. But the whole range of reference can be deepened and enlarged in the electronic books

So, neither multimedia, nor electronic books, nor internet-networks are a danger to the libraries, on the contrary they strengthen the bibliographic and informative work of the libraries and they may be considered as an evolution of libraries.

The appearance and the rapid development of computer networks and the computerization of the library process on the whole will result in the alteration of the reading zone and their connection with the stockroom and the internal working zone. As this procedure is occurring without any interruption and creating a dynamic which is getting more and more rapid, one of the most important results will be the fast changing of the centres of reading concentration, of the direction of movement and its spreading in the reading zone.

The use of the own or the libraries' computers by the reader is increasing. The card index will be steadily pushed to the periphery and lose its importance. The readers' motivation will change and also the priority of their self-chosen algorithm concerning the reading zone.

The use of computerized catalogues gains the character of bibliographic service in the reading zone. At the same time the computerized catalogues increasingly alter the bibliographic service into self-service. Thus the competent computer knowledge of the readers is becoming a more important factor.

The development of the information market and of the information service, the increasing competition and the increasing integration of libraries in this market force a more convenient service for the reader and that is the alteration of the librarian service into a library service. Solving this task is getting more important and at the same time more complicated within this difficult situation of the libraries due to their limited financial means.

The unforeseeable tendency of the development of libraries, their position in the information society under the condition of rapidly changing information connection structures, the role of multimedia and the socio-cultural situation demand a transformation of the reading zone. That includes its planning, new and rapidly changing technical equipment, higher demands on the acoustics, new equipment for the working places, their distribution in the reading zone and the setting-up of independent working places, their changing step by step into kind of a business service place with computer, scanner, connection to internet, a silent fax machine etc.

Many library users use the facility for furthering their own education and can get suggestions for intensifying their learning process. When libraries fulfil or even exceed the wishes of the user with their interior alterations, new technical equipment and the possibilities of literature searches, then - as past experience has shown us - the readers will use them even more frequently prefering them rather than a less service-friendly service institution.

It is only natural, that the needs of the reader are given the primary consideration when planning the library and its reading zone. At the same time it would be negligent not to consider the interests of the librarian, the influence of the architectural means on their ability to work, their motivation, their entire psychological state of mind. It would be wrong to presume that creating a reading zone which is favourable to the readers automatically creates pleasant conditions for the librarian. One should also consider at this point, that the librarian must spend the whole day at his work place and that dealing with many of the readers doesn't only produce positive emotions. At the same time, the reader judges the library by its librarian. And so the librarian is given a great responsibility, and the concern for a pleasant working atmosphere is not only a humanitary question, but also a pragmatic one.

This is how the relationship between the reader and the librarian looks. The technical conditions change but the basic relationships remain the same.

As far as the innovations are concerned:

About 34 years ago a well-known librarian, H. S. White, exclaimed: "To the barricades: The computers are coming." That is now history. In the meantime, the computers and computer networks feel at home in the library. We are now reaching a new stage – multimedia library, virtual library. That is the new future, but in a few years time this will be the routine in the everyday life of the library.

There will be new problems, new discussions and new conferences. C´est la vie ...

REFERENCES

1. see Heery, Rachel. Metadata Formats. December 1996. ukoln. ac. Uk / BIBLINK / wpl / dl / l.

2. Xu, Amanda. Metadata Conversion and the Library OPAC. 1998. web.mit.edu / waynej / www / xu.

3. B.I.T., 2-99.

9

CHAPTER

Academic Libraries in Transition: Current Trends, Future Prospects

Gregory A. Smith

Liberty University

ABSTRACT

Academic libraries are in transition because of changes in the context of higher education. Changes in the world of information are even more radical: the displacement of paper, the primacy of the search engine, the emergence of the digital lifestyle, and innovative patterns of scholarly communication. Decreasing reliance on local collections is transforming the library as a physical destination. Traditional measures of library success have begun to be replaced. Given the superiority of other information professionals' data management skills, the role of academic librarians will shift toward the enablement of learning. This environment of upheaval will pose both opportunities and challenges for academic librarians.

Authoring an article on the transformation of the academic library is a daunting task for at least three reasons. First, the literature on the subject is extensive, defying one's ability to gain a comprehensive understanding of the issues. Relevant information comes not only from the field of library science, but also from higher education, information technology, and other realms.

Authoring an article on the transformation of the academic library is a daunting task for at least three reasons. First, the literature on the subject is extensive, defying one's ability to gain a comprehensive understanding of the issues. Relevant information comes not only from the field of library science, but also from higher education, information technology, and other realms. S

Second, the radical changes taking place within the information environment seem to preclude any sort of accuracy in predicting the future of the academic library. According to Billings (2003), the notion that academic library development will proceed steadily along its current trajectory is unfounded; instead, unanticipated influences will dramatically alter the evolution of the academic library, frustrating any attempt to forecast with certainty its future state. The 2003 OCLC Environmental Scan thus advises in its introductory paragraphs, "Let us accept, then, that change is profound, accelerating, transforming and unpredictable. And let us also accept that, absent the talents of the Oracle of Delphi, any person or organization is unlikely to be able to make meaningful predictions that are helpful for charting directions for an indefinable future" (De Rosa & Dempsey, 2004, p. 1).

Third, current trends imply future prospects that are unsettling to many librarians, at the very least promising to bring about radical change in the nature of our duties, and possibly even threatening the future of our profession. According to De Rosa and Dempsey (2004), "There is a subdued sense of having lost control of what used to be a tidy, welldefined universe evident among those who work in this information environment" (p. 2). Crowley (2001) observes that "every so often a changing context so threatens a profession that the profession is forced to revisit issues thought settled long ago. At such times, the conventional wisdom tied to the accepted values of a given group often fails to provide answers to critical questions" (p. 569).

Taking these considerations into account, I must state that I am not comfortable with all of the prospective changes described in the article, nor am I convinced that all current trends will lead to a better future. However, I am confident that the trends to be discussed will impact academic libraries significantly and need to be responded to deliberately.

THE CONTEXT OF HIGHER EDUCATION

An academic library is not an end in itself, but a means to an end—namely, that of fulfilling a postsecondary institution's mission. "More than any other campus enterprise, the library symbolizes the distinct characteristics of the university and its mission across all disciplines: to develop the human intellect through teaching and learning and to contribute through research to the expanding body of human knowledge" (Wand, 2005, p. 4). Accordingly, it is most appropriate to begin this analysis of the future of academic libraries with a discussion of the context in which they operate.

Guskin & Marcy (2003) describe the historic assumptions of postsecondary education as follows:

Established organizational structures and processes for higher education were built to educate and support residential, traditionalaged students drawn from relatively homogeneous backgrounds whose prior education prepared them to attend college in a pre-technology-based learning environment. Faculty members were the primary instruments for imparting knowledge and skills, and individual classrooms remained the province of individual faculty members—who were also solely responsible for evaluating student performance. Completing a bachelor's degree in this setting is determined by the accumulation of individual classroom credits, assessed by discrete faculty members through the traditional grading process. (p. 18)

Such assumptions are no longer valid. Over the last few decades the environment has changed significantly, leading Dede (2005) to muse,

If civilization were to invent higher education today, rather than centuries ago, would we create campuses as they now exist, dominated by lecture rooms, libraries, and labs, with learning centered in fixed time blocks? I suspect instead we would design colleges and universities to distribute their activities broadly across geography and time, focusing on active construction of knowledge rather than assimilative incorporation of information. We now have the technological infrastructure to facilitate a reinvention of our historic approach, as well as promising models from many other sectors of civilization that have already reinvented their missions and organizational structures based on the capabilities of information technology. (p. 11)

Changes in the higher education context can first be seen in the demographics of the students who are attracted to our institutions. Today's university students are more ethnically and racially diverse than in the past. Increasing numbers attend college, many inadequately prepared for rigorous academic work. Non-traditional students—many with children— continue to seek out college degrees in order to enhance their economic opportunities. Students of typical college-going age—the millennials—share traits that are foreign to older generations:

high parental involvement, facility for technology-assisted communication, preference for group study, misunderstanding of intellectual property, and proclivity toward on-line transactions (Farrell, 2005).

The faculty is changing as well. Institutions are employing more part-time and adjunct professors, leading to a faster rotation of instructional personnel. Even the duties of the professor may be subject to change: "Reconsidering how faculty work in the context of new technologies and the roles of other campus professionals leads us to conceive of new roles for faculty members themselves. Instead of the standard lecture-discussion teaching format, faculty members may engage in a diverse array of roles, including mentor, intensive discussion leader, lecturer for short periods of time, and assessor of student mastery" (Guskin & Marcy, 2003, p. 17).

The economics of higher education represent an area of particular volatility. Costs are rising while government support is diminishing, shifting the burden of funding to students and extramural sources (Farrell, 2005, pp. 131-132). "Simply stated, costs are continuing to escalate beyond our ability to generate tuition and fund-raising revenues to cover them" (Guskin & Marcy, 2003, p. 12). Not surprisingly, "parents and students are approaching higher education with a new consumer mentality, a perspective prizing low cost, high quality, and convenience" (Crowley, 2001, p. 570). Left unchecked, these trends could spell disaster for the quality of higher education and, by implication, academic libraries.

Delivery methods have evolved significantly, due largely to the development of new information and communication technologies. Accordingly, the growth of distance education is eclipsing that of residential education. Guskin and Marcy (2003) argue that improving "the quality of faculty work life and student learning" necessitates "an educational delivery system that is built fundamentally upon the principle of recognizing and certifying student learning outcomes, wherever or however the learning occurs. The implicit assumption embedded in this approach is that the key productivity issue is not about how much faculty teach, but about how much students learn" (p. 16).

In the midst of these changes, colleges and universities have entered an era of greater accountability—to students and their parents, to government and other funding sources, and to accrediting bodies. The implications of this environment for academic libraries will be addressed in a separate section below.

Colleges and universities are, to a significant degree, in a state of crisis—caught between reverence for tradition and society's expectations of modernization. The strained context of higher education affects academic libraries profoundly. But another context, the information revolution, impacts them even more.

PATTERNS OF INFORMATION EXCHANGE

The Displacement of Paper

The word library derives from the Latin librarium, a derivative of liber, meaning "book." Viewed etymologically and historically, the library is a collection of bound paper volumes. Ironically, the developments of the past 25 years have rendered this characterization imprecise and, to a significant extent, inaccurate. Today's academic libraries disseminate information to their patrons very differently than they did a generation ago. In fact, it is possible to identify at least eight ways in which academic libraries have shifted away from amassing collections of paperbased monographs.

From analog to digital formats. Every major category of information-bearing media—text, image, audio, and video—has become increasingly available in digital form over the last 10 years. Digital distribution already has a proven track record among serial publications, reference sources, and government documents. While the economics of copyright and the limitations of display technology have delayed the adoption of e-books, use statistics and user feedback at the author's library suggest that college students are accepting them with increasing enthusiasm.

From books to journals and other media. While books continue to be published in abundance, the center of discourse in many academic disciplines has shifted to the medium of the scholarly journal. In addition, the volume of non-textual media available for acquisition or access is expanding. According to Hazen (2000), "Print publications show no sign of disappearing . . . . Nonetheless, print collections are losing their one time preeminence relative to library holdings in other media. Shared resources and remote digital products are likewise reducing libraries' reliance on in-house collections. Book-based bibliographers are already pretty much obsolete in some natural sciences, and those servicing many other fields are eventually likely to follow suit" (p. 838).

From highly accessible on-site storage to compact storage, whether on or off site. Libraries have historically allocated prime space to the storage of their physical collections, sometimes at the expense of reading and study space. Given the shift toward on-line research over the last decade, it comes as no surprise that many academic libraries are replacing their high-availability stacks with denser forms of storage, whether on the premises of the library or elsewhere (Freeman, 2005, p. 8; Shill & Tonner, 2004, pp. 140-142).

From local storage to remote access. An increasing proportion of the information purveyed by the typical academic library is not housed within the library building, but is accessed from off-site servers. Symposium 2010 participants envisioned a future in which "much content is leased by the library which links the user to the information stored at remote locations. Vendors own some of it and some is owned collectively by libraries cooperating in consortia where the archiving of the retrospective cultural heritage is the responsibility of a few libraries who agree to make it available to others in perpetuity" (Wand, 2005, p. 5).

From local ownership to subscription-based access. As noted above, most libraries have relinquished responsibility for data storage to a significant extent. In addition, they often access remote databases via term-based licensing, an economic model that is antithetical to the historic concept of local library ownership

From selection of individual items to selection of resources in the aggregate. As libraries have transitioned from developing local collections of books and media to the licensing of digital content available from vendors, they have often agreed to select predefined aggregations of full-text journals, ebooks, on-line reference works, and other resources.

To varying degrees, professional librarians have conceded one of their primary functions: the selection of individual book, journal, and media titles to meet the specific needs of their users.

From library-specific collection development to group-based resource-sharing. Until recently, academic libraries made collection development decisions with a fair degree of autonomy. The most cooperative among us negotiated with other institutions to build areas of collection strength that were complementary rather than duplicative. However, the emergence of numerous library consortia has led to the sharing of resources among groups of libraries. Such sharing is most notably seen in consortial database licensing, but has also been implemented in the form of depository libraries serving multiple institutions.

From active acquisition of grey literature to free access via the Web. According to Mathews (2004), grey literature "is commonly defined as any documentary material that is not commercially published and is typically composed of technical reports, working papers, business documents, and conference proceedings" (¶ 1). It is produced by organizations whose primary role is something other than publishing (Weintraub, n.d., ¶ 1). In the preWeb era such literature was difficult to identify and acquire, so academic libraries either neglected it or expended significant effort to collect it. However, many non-profit organizations, educational institutions, government bodies, and other entities now disseminate information to interested parties via freely accessible Web sites. As a result, researchers can access the content as their needs dictate, and libraries play a less vital role in making it accessible—a significant shift (Weintraub, n.d.; Mathews, 2004).

Taken collectively, these changes argue convincingly that academic libraries are not what they were in recent memory. After a long-standing reign, paper has been displaced as the primary means of information exchange. Of course, libraries will continue to preserve information on paper for the foreseeable future. Christian college libraries may do so longer than their secular counterparts if Christian literature is slow to appear in digital form—whether because certain publishers cling fervently to paperbased distribution, or because mass digitization projects fail to target many Christian publications. Nevertheless, while libraries will provide access to legacy paper collections for a long time to come, digital distribution will increasingly dominate the information industry.

The academic community is experiencing the fulfillment of a phenomenon that Clifford Lynch anticipated nearly a decade ago: "Now that we are starting to see, in libraries, full-text showing up online, I think we are very shortly going to cross a sort of a critical mass boundary where those publications that are not instantly available in fulltext will become kind of second-rate in a sense, not because their quality is low, but just because people will prefer the accessibility of things they can get right away. They will become much less visible to the reader community" (Educom Review Staff, 1997). The accuracy of Lynch's observation is confirmed even in this article's bibliography, which is comprised mostly of journal articles and other sources available in electronic form, some of which were chosen over less conveniently accessible print materials.

The Primacy of the Search Engine

Related to the displacement of paper is the emergence of on-line search engines that, in about a decade, have replaced library catalogs and other bibliographic tools as the most common places to begin looking for information. With the launch of Google Scholar in November 2004, the search engine industry made a concerted effort to penetrate the world of scholarly research.

One has to look no further than trends in the advertising industry in order to understand the significance of on-line search in today's economy. Gross reported in mid-2005 that television, magazine, and newspaper advertising were in decline, while Internet advertising—tied largely to search engines— was expected to increase by 15% over the previous year (2).

Search engines are beginning to move beyond the scope of static, freely accessible Web pages into what is sometimes referred to as the "deep Web." The deep Web is a vast array of information located in on-line databases whose content is only served up in response to a searcher's query (e.g., WorldCat records). Formerly invisible to search engines, some deep Web sites are now searchable along with the

Web's static content. Search engines increasingly provide links to deep Web content that is available for a fee (e.g., from document delivery services). Google and its competitors have established themselves within the search market and are unlikely to be challenged by database vendors traditionally associated with library research (EBSCO, ProQuest, Gale, etc.). Rather, several library-oriented companies have already announced partnerships with Google. This trend is likely to continue, enabling standard Internet search engines to provide pathways into proprietary database content.

Within the last 18 months a number of worldclass technology companies and libraries have announced their intent to launch several ambitious digitization projects (e.g., Open Content Alliance, Google Book Search, World Digital Library). The indexing of the resulting files by search engines will only serve to strengthen their identity as the tool of choice for research. Furthermore, libraries will face increasing competition from innovative corporations who provide digital information directly to the consumer, whether via subscription (e.g., Questia) or on a pay-per-page basis (e.g., Amazon Pages).

All of this serves to indicate that the search engines are a force that libraries (yes, even academic libraries) must acknowledge. Librarians seem to be waking up to this fact, as evidenced by the coverage Google received in American Libraries during 2005. Of course, librarians' opinions of search engines run a wide gamut. On the one hand, Caufield (2005) argues that Google's success is partially attributable to its adoption of traditional library values such as the facilitation of access to information—through simple interfaces, with relative lack of bias vis-à-vis content. On the other hand, authors such as Bell (2005) and Mann (2005) view Google as antithetical to the intelligent subject analysis and information literacy instruction that are arguably essential to librarianship. And Stewart (2006) argues that searching the full text of books may prove detrimental to theological scholarship, which requires deep, contextual reflection.

Libraries' potential responses to Google and other search engines are analogous to the various ways that professing Christians relate to their cultural milieu, as construed in Niebuhr's (1951) Christ and Culture.

- The library above Google: ignore Google, even if this results in irrelevance to users
- Google as the library: concede to Google's identity as the ultimate library
- The library against Google: criticize Google for its shortcomings and oppose it for distracting users from the superior quality of the library
- The library and Google as awkward relatives: coexist in the same space with Google, yet never resolve the tension between it and the library
- The library the transformer of Google: influence Google to make it more library-oriented; make authoritative resources searchable through Google

The Emergence of the Digital Lifestyle

The continuous proliferation of digital content described in the previous section will doubtless affect society in ways we can hardly anticipate.

> It goes without saying that the economic impact of the e-book on publishers and booksellers will be dramatic . . . . But I'm more interested in how the e-book will affect the way we read—and write. New technologies, after all, change art, often in profound and unpredictable ways. I doubt the inventor of the electric guitar foresaw Jimi Hendrix, any more than Thomas Edison foresaw chick flicks. The only thing of which you can be certain is that the existence of the e-book will cause the authors of the 21st century to go about their business very differently than did their 20th-century predecessors. (Teachout, 2006, 7)

Digital networks, both wired and wireless, now make it possible for citizens of technologically advanced nations to conduct many life functions— including work, entertainment, education, and social relationships—virtually anywhere. As a result, traditional means of interacting with people and ideas, including some functions historically carried out within library buildings, are subject to transformation (Mitchell, 2005). Implications of these changes for the library as place are significant, and will be addressed in a separate section below. This section will discuss the emergence of the digital lifestyle, a way of life that is made possible by the abundance of digital media, the pervasiveness of advanced communication networks, and the emergence of increasingly portable electronic devices (Barna Group, 2006).

Different people—and groups thereof— participate in the digital lifestyle to varying degrees. Citizens of highly developed nations increasingly conduct life through the mediation of information technology. Download services, blogs, instant messaging, podcasting, on-line financial transactions, video on demand, Internet-based telephone service, RSS, PDAs, computer desktop search software, and text messaging are all manifestations of the digital lifestyle. High school and college students interact more freely with such technologies than most librarians—despite our training as information professionals. In fact, the technologies are largely transparent to them; they are simply a part of how their world works.

The Horizon Report (2006) projects that social computing and personal broadcasting will achieve significant impact on higher education within one year or less; that high-function mobile phones and educational gaming will achieve this status in two to three years; and that within four or five years, augmented reality, enhanced visualization, and context-aware systems will shape teaching and learning at the postsecondary level. The fact is that technological innovation is accelerating. The students who will come to our campuses—or perhaps only interact with us from a distance—within the next five years will expect their college experience to be digitally-enabled.

Whether we like it or not, our current and prospective students are accustomed to using simple, intuitive Web interfaces to secure information for themselves (De Rosa & Dempsey, 2004, p. 4). Academic libraries are unlikely to defy this trend. Our patrons expect to retrieve and manipulate information without the mediation of the library staff (Campbell, 2006, pp. 22, 24). In light of our users' behavior, we should aim to develop a Web presence that facilitates self-service. To do so we will need to partner with highly innovative software developers and on-line service providers. Our suite of tools may include the following:

- A knowledge base that allows users to search and browse for answers to frequently asked questions about the library

- Information literacy instruction disseminated through blogs, podcasts, Web-based tutorials, and other popular distribution channels
- Research advisory tools that mimic some of the interaction that takes place in a reference interview
- Federated search functions that provide a bird'seye view of the databases that may provide the answer to a user's query
- Link resolvers that make the path from citation to full text as direct as possible
- Library catalogs that incorporate the best features of on-line bookseller sites and search engines
- Catalog enrichment services that provide a rich array of information—tables of contents, reviews, cover images, and more—about materials on our shelves
- Portal and alert services that actively push content to end users based on past activity or stated preferences
- Virtual reference services that are visible where users conduct searches and available at times that are convenient to them, and that pose minimal technical obstacles for use
- On-line citation services that assist patrons in building bibliographies

Above all, it is vital that our on-line resources and services be highly integrated and transparent to end users. Participants in Symposium 2010 expressed their vision for this as follows: "The scholar, student, [and] administrator become the central focus as the library becomes less visible and more integrated into the infrastructure of the enterprise" (Wand, 2005, p. 2).

Implementing systems and services such as those described above will not be easy. Success will be a process, not a destination. It is helpful to observe that "in most cases the effective application of information technologies for competitive operational advantage requires that the business processes be reengineered" (Cortez, Dutta, & Kazlauskas, 2004, p. 132). In short, this means that we will have to start thinking differently about the library enterprise if we are to harness the value of emerging technologies and become an integral part of the digital lifestyle.

The Comparative Roles of Libraries and Publishers

Stinson (2006) aptly observed that "publishing is a sibling, if not a parent, of librarianship" (p. 14). In this author's view, the relationship was, until recently, quite parental, with libraries essentially deriving their collections from publishers much as children inherit traits and learn habits from their parents. However, emerging models for the publication and distribution of scholarly information may foster between libraries and publishers the sort of competition that often characterizes sibling relations.

Historically, a scholarly book or article's path from author to library patron was fairly easy to trace: author, publisher, distributor, acquiring library, reader. This flow will probably become much more diversified as the economics of digital information take shape. Possible alternative paths include:

- Author, publisher, database aggregator, subscribing library, reader

- Author, publisher (functioning as database vendor), subscribing library, reader
- Author, publisher, database aggregator, reader
- Author, sponsoring library (as publisher), search engine, reader
- Author, publisher (distributing directly to end user), reader

According to De Rosa and Dempsey (2004),

It is clear that a new ecology and a new economy for scholarly materials are being formed. In the past, the flow of research and learning outputs traveled through formal, linear publishing mechanisms. We are seeing the emergence of a variety of repository frameworks, metadata aggregation services, and richer content interconnection and repurposing that are changing how we think about data and its uses. The library has the opportunity to take a leadership role in developing policies and programs that contribute to a coherent, institution-wide knowledge management system. (p. 11)

Academic libraries will conceivably emerge from this unstable environment with two new roles. First, our organizations will quite likely take charge of digital rights management on behalf of our institutions' information resources (Campbell, 2006, pp. 24, 26; The Horizon Report, 2006, p. 4). Second, many of our libraries may emerge as hosts of institutional repositories that bypass some of the functions carried out historically by scholarly publishers (Campbell, 2006, pp. 26, 28; Wand, 2005, p. 3). Christian college libraries may bear a particular responsibility to facilitate the digital preservation and distribution of retrospective Christian content.

Nevertheless, it is prudent to note that publishers and distributors are seizing (and will surely continue to seize) opportunities to disseminate content directly to end users—without the intermediation of libraries. Thus, both publishers and libraries will find it necessary to assess and defend their roles in the face of new competition, and it is conceivable that the distinctions between the two will be blurred.

THE ROLES OF THE LIBRARY AS PLACE

Library buildings historically fulfilled three primary functions for their patrons: First, they stored collections of books and other information-bearing materials. Second, they provided space for patrons to read, study, and meet with each other. Third, they allowed researchers to consult with staff members concerning their information needs. As long as the library remained essentially a collection of material objects, it would have been incoherent to question its importance as a physical destination. However, given the progressive migration of library resources and services from physical space to cyberspace, it has become quite appropriate to talk about the future prospects of the library as place.

Over the course of the last two decades computer networks have revolutionized the way that libraries meet the needs of their patrons. Reading and research are becoming progressively less dependent on users' location. To a significant extent it is no longer necessary to visit library buildings in order to retrieve information resources. Given such trends, one wonders whether there is much of a future for library buildings, particularly on college and university campuses.

Crowley (2001) regards this situation as a serious threat. In his view, academic libraries' successful deployment of on-line database resources has led, in many cases, to reduced foot traffic within the library facility. Given the strained financial condition of the higher education industry, a climate seems to be emerging in which some college and university leaders may find it reasonable to diminish or eliminate physical libraries (pp. 572-573). Campbell (2006), for his part, focuses on the incongruity of allocating prime space to the storage of low-use collections; he predicts that such space will be reassigned, though not necessarily to functions typically associated with academic libraries (p. 20).

Notwithstanding these concerns, there is cause for hope. Many campuses continue to construct, renovate, or expand library facilities. The number of new academic library construction projects reported annually in the December issue of Library Journal (LJ) has remained fairly constant since 2002. Perhaps the most intriguing aspect of the LJ articles is the apparent trend toward the integration of library services with other campus functions, both academic and non-academic. Innovation is reflected even in the names conferred on some of the new facilities, as shown below.

- Names emphasizing the library's roles vis-à-vis learning: Academic Resource Center; Learning Resource Center (2); Library and Learning Center (2); Student Learning Center
- Names emphasizing the connection between the library and information technology: Center for Library & Information Resources; Digital Library & Learning Resource Center; Information Commons; Library & Computer Commons; Library, Research, & Information Technology Center

- Names suggesting a desire to make the library a high-profile destination: Information & Alumni Center; Library & Student Center
- Name indicating a commitment to long-term preservation of physical collections: Library Depository/Retrieval Facility

Shill and Tonner (2004) report that "80 percent of the libraries completing a major space improvement project between 1995 and 2002 experienced greater facility usage in 2001-2002 than they did in a preproject baseline year . . . . The median change in postoccupancy usage was a 37.4 percent increase" (p. 148). They conclude that their research provides clear, empirical evidence that students can and will use a comfortable, well-equipped library, even with remote access to many electronic databases and the Internet available. This is an important conclusion because it suggests that a discerning investment in library facility improvements—whether a new library or improvements to existing space—will attract students to a specialized physical place designed to provide research and study space, teach information literacy skills, expose students to recorded knowledge in both print and electronic formats, and make "information experts" readily accessible. (p. 149)

Current trends suggest that the academic library's role as information warehouse will gradually decrease in importance. Last year the University of Texas at Austin moved most of the 90,000 volumes in its undergraduate library to other sites around campus (Blumenthal, 2005; Flawn Academic Center, 2006). The newest campus in the University of California system, located in Merced, aims to maintain a collection of only 250,000 volumes—by historical and comparative standards, a slim figure for a research library (Carlson, 2005). Diminishing reliance on physical collections will likely result in more widespread deployment of compact shelving, maintenance of zero-growth collections, removal of little used print collections to remote storage facilities, and increased use of interlibrary loan and document delivery services.

Under ideal circumstances, space formerly designated for existing or planned collections will be reallocated for patron use. According to Symposium 2010 participants, "The [future] library is a welcoming, comfortable, functional, meeting place. The size of the onsite book collection is relatively stable and the traditional predominant pressure to accommodate its growth is replaced by reconfiguring space for the learner" (Wand, 2005, p. 3). According to one prominent library architect, this will involve a radical shift: "'With the libraries of the past,' explains [Geoffrey] Freeman, 'you projected the rate of acquisition of a collection for 20 years. It always expanded at the expense of the user function. It's just the opposite now. Now you project it out to zero growth. You design around the user and expand at the expense of the collection'" (Allis, 2005, ¶ 13).

Focusing library space design around users will require academic communities to adapt their thinking to a generation of students whose skills, habits, and preferences differ from those of years gone by. As Farrell (2005) notes,

Librarians have seen a trend toward group study and projects within academic libraries. Teamwork is a hallmark for millennials and they thrive in a group environment. . . . This will require spaces within libraries that support group interactions and technology. Librarians in addition to "libraries as a virtual resource" must focus on the "library as a place." The mix of remote and onsite services will be a challenge as librarians have been focusing on electronic access to collections. (p. 132)

Freeman (2005) succinctly summarizes this new view of the academic library as place: "The library's primary role is to advance and enrich the student's educational experience; however, by cutting across all disciplines and functions, the library also serves a significant social role. It is a place where people come together on levels and in ways that they might not in the residence hall, classroom, or off-campus location" (p. 6).

CRITERIA FOR MEASURING LIBRARY PERFORMANCE

As this article has already documented, the academic library landscape has changed significantly—even within the last five years. Not surprisingly, the traditional measures of an academic library's success—perhaps most clearly embodied in the standards of accrediting and professional bodies—have begun to be replaced.

Twenty years ago, the marks of an outstanding academic library were fairly clear: impressive physical facilities, massive collections of books and subscriptions (on paper or microform, of course), seating for a significant proportion of the student body, a budget based on the institution's educational and general expenditures, and services provided by specialists in the disciplines of concern to the parent institution. Today, such measures are of lesser consequence, and others have arisen to significant prominence (Kyrillidou, 2004). Simply put, the application of technology to scholarship and libraries has led to a situation in which there are numerous ways to approach the satisfaction of information needs. Accordingly, accreditation standards for libraries are less prescriptive than they used to be.

To be clear, the displacement of the older standards was not accompanied by a decrease in the accountability of academic libraries. In fact, following trends in government and virtually every other segment of higher education, academic libraries have come under increased scrutiny in recent years. Patrons and regulatory agencies now expect academic libraries to demonstrate their value via various modes of assessment. A library's value is increasingly measured in relation to its stated mission, especially as that mission correlates with users' satisfaction and learning (De Rosa & Dempsey, 2004, p. 7; GratchLindauer, 2002). Indeed, creati

Indeed, creating experiential value for the patron is overwhelming all other criteria of success. In today's economy, the campus library faces a significant amount of competition from other players within the information marketplace. Not surprisingly, actual and prospective patrons tend to measure the library's quality and efficiency by comparing its facilities, resources, and services to those of its competitors: Barnes & Noble superstores, the various Google services, Amazon, and Questia, to name a handful.

As a case in point, Coffman (1998) compares the operation of bookseller chain superstores and branch public libraries, outlining both similarities and differences. He argues that bookstores operate more efficiently (by a margin of about 30%) by hiring less expensive personnel, offering a lower standard of information service, and spending less effort to catalog and classify the material on their shelves. At the same time, they manage to provide surroundings and service hours that exceed the comfort and convenience of many libraries. Clearly, the aims of bookstores and libraries overlap, at least to some extent, and users are bound to apply the standards of the former when evaluating the latter.

Unfortunately, as Crowley (2001) notes, institutional decision-makers may take little account of the virtues of academic libraries when comparing them to their various competitors.

For an academic librarian, the rise of what are increasingly seen as more or less acceptable electronic alternatives to her or his professional services should be a very strong stimulus to identifying where librarians really rank in their own educational contexts. Here, it must be stressed that librarians make a crucial mistake if they believe that boards of trustees, presidents, and research/teaching faculty only support alternatives to university programs that are better than, or at least as good as, the originals. In reality, substitute services do not have to offer better or comparable quality. Replacement services only have to be "good enough to get the job done," according to the operative definitions of those who make significant college and university decisions on resource allocation. (p. 566)

Nevertheless, there is much to be said for usercentered academic libraries. If use of our libraries is suffering due to public perception, we need to do a better job of marketing what we offer and why—to students, professors, administrators, and trustees. But we cannot expect users to adapt extensively to our traditions. Rather, we must update what we do to conform to current societal expectations. Even more invasively, we need to think about conceding some of our historic roles and adopting others in their place. According to Anderson (2006), the informationseeking behavior of actual and potential library users is changing significantly. "Like water, they will follow the path of least resistance" (p. 1). Libraries are increasingly measured by the standards of popular Web-based tools. If libraries neglect (or worse, try to override) patron preferences, "they'll ultimately go where they want and we'll just get hurt" (p. 1). Accordingly, libraries must change—or fade into obsolescence.

LIBRARIANSHIP AMONG THE PROFESSIONS

The preceding pages have outlined various trends that will impact academic libraries and librarians in the foreseeable future. The picture sketched here is not always inspiring; the library profession is in jeopardy, threatened by competition from a variety of quarters. Appropriately, this final section will attempt to explore whether there remains a long-range future for academic librarians, and if so, what roles we may play in it.

Required by an overwhelming majority of library jobs for more than 30 years, the master of library science (M.L.S.) degree is the unquestioned historic credential for admission to the library profession (Lynch & Smith, 2001, p. 414). However, until recently, the content of a typical M.L.S. degree did little to prepare a librarian for the technical duties that are increasingly characteristic of library jobs. According to Higa et al. (2005), "Analyses of job trends indicate that advanced computer skills, proficiency in Web-based resources and services, and the ability to be creative and inventive in an online environment are prerequisites for many library positions. Hiring new librarians with these skills and improving already-employed librarians' computerbased abilities is of paramount importance in an electronic environment" (p. 43).

Cortez, Dutta, and Kazlauskas (2004) press the case even further:

In various types of libraries, staff and organizational structures are changing. In terms of staffing trends there is less need and emphasis on in-house technical specialties— acquisitions, cataloging, and processing— because these services are either being outsourced or assigned to paraprofessionals, particularly in many academic and school libraries. . . . The resulting role for the information professional or librarian aligns more closely with the knowledge and skills in technology management, business operations, and interpersonal skills. (pp. 134-135)

These authors go on to specify four categories of knowledge, competencies, and skills for library and information science professionals: technical (e.g., tools, processes, subject expertise), administrative (e.g., leadership, project management), social (e.g., communication, teamwork, problem-solving), and system (e.g., development methodologies, business planning) (pp. 139-140).

The boundaries of librarianship are more arbitrarily and rigidly defined than those of other information professions. The M.L.S. establishes a baseline of knowledge and skill among those who hold it; however, it often seems to outline limits of potential innovation. Deiss (2004) astutely observes that the relative maturity of many libraries entrenches them in traditions that tend to stifle needed innovations (pp. 23-24). Furthermore, our selfperception as experts can deter us from experimenting with new areas of discovery (p. 25). It is possible that, in the face of rapid technological change, our colleagues in the newer information professions are less convinced of their own expertness and more willing to engage in creatively playful behaviors that lead to innovation

Crowley (2001) notes that "the revolution in electronic information brought about in large part by the academic library community has contributed to career uncertainty for its originators" (p. 581). This is to say that, through our successful deployment of online resources and services, we have unwittingly initiated processes that have gone much further than we intended. Whereas in years past we were visionaries whose innovations pushed patrons to adopt new information technologies, social forces are now forcing us to escape our own comfort zones.

As one compares the skill sets of librarians with those of more technically-oriented information professionals, it is painfully obvious that others can manage raw data more efficiently than we can. According to Campbell (2006),

The skills needed to work with metadata, IRs, and other similar sources are much more highly technical than those possessed by most of today's academic librarians. In bringing such questions [about the future of academic libraries] forward, therefore, librarians must understand the stakes involved. . . . Given the events of the past decade, academic librarians perhaps know better than anyone else that the institutions they manage—and their own roles—may face extinction over the next decade. (p. 28)

Marcum (2003) summarizes the entries submitted in connection with an essay contest concerning the academic library in 2012. According to one contestant, the technological future anticipated for the academic library can only be realized through "reliance on outsourcing or a serious revitalization [of] the library profession involving the development of new roles and improved status for librariantechnologists" (¶ 7). Following this logic, the academic library profession can either be expected to shrink (as more functions are outsourced) or to undergo profound transformation (as current members acquire new technical knowledge and skills, and as tech-savvy members are recruited to replace those who leave the profession). But this may be an oversimplification. Though our data management skills are not state-of-the-art, we may be able to handle information more intelligently than others— with knowledge of the disciplinary context and the capacity to facilitate the acquisition of knowledge.

In light of the above, academic librarians' value proposition will increasingly shift from the direct management of information toward the enablement of learning. According to Baker (2006), "Placing learning as the organizing principle for all that is done in the academic library is qualitatively different from simply understanding that it serves some educational role. It suggests that all the roles played by the academic library ought to be functions of the overarching aim to meet each one who comes in such a way that it facilitates an experience that engages them in authentic learning" (p. 8). Recently published regional accreditation standards seem to anticipate this. References to library collections have yielded to the language of access, reflecting the displacement of librarians as sole custodians of scholarly information. However, information literacy instruction has risen to greater prominence and there is a clearer sense of the need for an information-rich learning environment, reflecting areas where libraries may contribute to learning outcomes (Gratch-Lindauer, 2002, p. 16).

The changes anticipated here may well serve to decrease the hegemony of the M.L.S. as the standard credential for professional library service. Though library schools have already modified (and will surely continue to modify) their curriculum to meet evolving needs, the diversification of academic library roles may well preclude them from continuing to serve as nearly exclusive providers of education for librarianship. It is difficult to imagine that many schools of library and information science will be capable of providing adequate instruction across the broad range of disciplines that will likely inform librarianship in the future: library science, computer science, information science, communication studies, education, and management. Accordingly, academic libraries will presumably become more open toward the idea of hiring candidates with degrees other than the M.L.S., especially if they hold the Ph.D. (Berry, 2003; Crowley, 2001, pp. 580-581)

Given the likelihood that the role of the academic librarian will be transformed in some significant ways in the coming years, and that this will presumably diversify the range of credentials represented within the profession, it is difficult to predict whether the traditional values of librarianship will remain intact. How many of Gorman's (2000) core values—stewardship, service, intellectual freedom, rationalism, literacy and learning, equity of access, privacy, and democracy—are likely to be upheld in an environment where information resources are increasingly managed by commercial entities? Will those who are admitted into the library profession from disciplinary traditions other than library science imbibe the historic commitments of librarians? It is difficult to hazard a prediction.

On a related yet distinct note, what are the prospects for future Christian influence within the library profession? The pursuit of a "deep librarianship," to borrow Richardson's (1992) phrase, has in the past been frustrated by the influence of radical libertarian thought and by the all-too-frequent attention of professional organizations to political causes that have little to do with the success of libraries (Durant, 2005). There is definitely an ongoing need for organizations such as the Association of Christian Librarians to provide spiritual orientation within the profession—as a ministry to Christians working in religious and secular contexts, and as a testimony to librarians who are opposed to the faith.

If, as described above, the future of academic librarianship will tend more towards pedagogy than information management, librarians serving Christian institutions will have particular opportunities to aid students and faculty in the integration of faith, life, and learning (Smith, 2000/2002a). Librarians serving in non-Christian settings will doubtless need much support as they seek to share their faith responsibly and advocate for the collection of Christian materials within their libraries (Davis, 1992; Davis & Tucker, 1993/2003). Those of us who wish to express our interest in the implementation of family-friendly policies within public and school libraries will probably find sympathetic voices among our Christian librarian colleagues. Some of us need to work on articulating a philosophy of librarianship consistent with biblical theism, a framework that will help us all to rationalize our identity as Christian librarians (Smith, 2002b; Waller, 1977). Many of us may find that networking with other Christian librarians motivates and empowers us to assist in the development of Christian libraries overseas (Abernathy & Gill, 2003). Finally, all of us will surely benefit from mutual Christian encouragement as we continue to navigate an environment of rapid change

CONCLUSION

In light of the trends discussed in this article, what conclusions can we reach concerning the future of academic libraries? First, the academic library is not an end in itself. For too long we have acted as if libraries hold the same status as motherhood and apple pie. We must now wake up to the fact that libraries are in jeopardy—that we have to prove our value—in a market that makes information resources and services available, without our intermediation, to our intended patrons.

Second, if academic libraries are to remain a vital component of their parent institutions, academic librarians must understand the stresses that higher education is facing. We have a strong history of providing support for teaching and learning activities, yet postsecondary institutions' academic support needs are changing. If we can discover what academic support needs are currently going unmet and adapt to meet them, we will likely ensure our collective viability (though modifying our profession from its traditional form). If we perceive ourselves as being in the information management business, we will probably lose out to other players in the market. However, if we identify our past involvement in information management as an expression of our support for teaching and learning, we will find that we are better poised to assume other roles in the academic support arena.

Third, the notion of a library as a collection of locally owned materials—especially books—is out of touch with reality. We should expect continued migration towards a digital information environment. Information is being packaged in a more diverse range of containers than ever before. It is being delivered to end users with decreasing regard for their location, often without direct human mediation. We can expect continued digitization of retrospective content as well as intensified competition from other information providers.

Fourth, individual academic library facilities will remain the heart of their respective institutions only to the extent that they intertwine themselves with learning. Core learning activities may include classroom instruction, lab-based instruction, computer-based research, private study, interaction in virtual communities, and collaboration between faculty, staff, and students. Future library construction and renovation projects should emphasize flexible design, give user needs and preferences primary consideration, integrate a variety of current and emerging technologies, and lead to increased partnership with other stakeholders in the institution's teaching and learning mission.

Fifth, academic libraries will continue to outsource or concede their information management functions to non-library entities. To the extent that they develop programming to support local educational needs, they will become increasingly unique. Absent the emphasis on storage of local collections, it will become progressively more difficult to define the essence of the academic library. Accordingly, assessment of library performance will come to be defined even more in terms of its contribution to locally defined learning objectives.

Sixth, libraries that fail to innovate will die a slow death. Factors contributing to their demise may include end users' overwhelming orientation toward network-based information resources; the proliferation of on-line resources and services that bypass libraries; library employees' resistance to change; the myopia of campus leaders who perceive no need to invest in libraries in an environment of abundant Web-based information; and the preoccupation of professional organizations with political issues and other matters less than central to the success of the library.

Is there a future for academic librarians? My answer is a qualified "yes." Moving beyond mere survival to a renewed position of centrality will require us to change in fundamental ways. We must be open to change in our organizational mission, our individual job descriptions, the credentials required for work in libraries, patterns of library facility use, and expectations of a highly stable work environment. If we listen attentively to our institutional communities, we will discover that there are many areas of untapped opportunity for academic librarians and others who are passionate about serving the noble cause of higher learning.

REFERENCES

1. Abernathy, W. F., & Gill, K. D. (2003). Partnership in library development: The mission focus of Christian librarians. In G. A. Smith (Ed.), Christian librarianship: Essays on the integration of faith and profession (pp. 192-200). Jefferson, NC: McFarland.

2. Allis, S. (2005, July 31). Architect of learning: The reinvention of the college library. The Boston Globe. Retrieved February 13, 2006, from http:// www.boston.com/news/education/higher/ articles/ 2005/07/31/architect_of_learning/

3. Anderson, R. (2006, February). Always a river, sometimes a library. Address presented at the Soaring to Excellence 2006 virtual conference. Retrieved February 22, 2006, from http://www.dupagepress .com/COD/uploads/media /Handout_Packet_2_01 .pdf

4. Baker, S. L. (2006). Sustaining the cultural icon through purposeful renewal. Unpublished manuscript, Union University, Jackson, TN.

5. The Barna Group. (2006, February 7). Americans' on-thego lifestyles and entertainment appetites fuel increasing reliance upon technology. Retrieved February 8, 2006, from http://www.barna.org/ FlexPage.aspx? Page=BarnaUpdateNarrow& BarnaUpdateID=217

6. Bell, S. J. (2005, October). Submit or resist: Librarianship in the age of Google. American Libraries, 36, 68- 71.

7. Berry, J. N., III. (2003, November 1). But don't call 'em librarians. Library Journal, 128, 34-36. Retrieved February

13, 2006, from http://www .libraryjournal.com/ article/CA329316.html

8. Billings, H. (2003). The wild-card academic library in 2013. College & Research Libraries, 64, 105- 109. Retrieved February 16, 2006, from http:// www.ala.org/ala/acrl/acrlpubs/crljournal/ backissues 2003b/march03/billings.PDF

9. Blumenthal, R. (2005, May 14). College libraries set aside books in a digital age. The New York Times. Retrieved June 5, 2005, from Factiva database.

10. Campbell, J. D. (2006, January/February). Changing a cultural icon: The academic library as a virtual destination. EDUCAUSE Review, 41, 16-30. Retrieved February 3, 2006, from http://www .educause.edu/ir/library /pdf/erm0610.pdf

11. Carlson, S. (2005, April 1). Shelve under "e" for electronic. The Chronicle of Higher Education. Retrieved March 20, 2006, from LexisNexis

12. Caufield, J. (2005). Where did Google get its value? portal: Libraries and the Academy, 5, 555-572.

13. Coffman, S. (1998, March). What if you ran your library like a bookstore? American Libraries, 29, 40-46. Retrieved February 26, 2006, from Academic Search Premier database.

14. Cortez, E. M., Dutta, S. K., & Kazlauskas, E. J. (2004). What the library and information professional can learn from the information technology and project management knowledge areas. portal: Libraries and the Academy, 2, 131-144.

15. Crowley, B. (2001). Tacit knowledge, tacit ignorance, and the future of academic librarianship. College & Research Libraries, 62, 565-584. Retrieved February 16, 2006, from http://www.ala.org/ala/ acrl/acrlpubs/crljournal/ backissues2001b/ november01/crowley.pdf

16. Davis, D. G., Jr. (1992). "At all times, and in all places": Christian witness in international librarianship. Christian Librarian [U.K.], 16, 7-14.

17. Davis, D. G., Jr., & Tucker, J. M. (2003). The Master we serve: The call of the Christian librarian to the secular workplace. In G. A. Smith (Ed.), Christian librarianship: Essays on the integration of faith and profession (pp. 40-47). Jefferson, NC: McFarland. (Reprinted from Christian Librarian [U.K.], 17, 1993, 17-25)

18. De Rosa, C., & Dempsey, L. (2004). The 2003 OCLC environmental scan: Pattern recognition: Executive summary. Dublin, OH: OCLC. Retrieved December 14, 2005, from http://www .oclc.org/membership escan/downloads/ escansummary_en.pdf

19. Dede, C. (2005). Planning for neomillennial learning styles. EDUCAUSE Quarterly, 28(1), 7-12. Retrieved February 13, 2006, from http://www .educause.edu/ir/library /pdf/eqm0511.pdf

20. Deiss, K. J. (2004). Innovation and strategy: Risk and choice in shaping user-centered libraries. Library Trends, 53, 17-32. Retrieved February 26, 2006, from Academic Search Premier database.

21. Durant, D. (2005, September 30). The loneliness of a conservative librarian. The Chronicle of Higher Education.

Retrieved September 28, 2005, from http://chronicle.com/free/v52/i06/06b01201.htm

22. Educom Review Staff. (1997). Networked information: Finding what's out there: Clifford A. Lynch interview. Educom Review, 32(6). Retrieved February 13, 2006, from http://www.educause .edu/apps/er/review/review Articles/32642.html

23. Farrell, M. (2005, April). Trends in higher education: How will they impact academic libraries? Paper presented at the 2005 Conference of the Association of College & Research Libraries, Minneapolis, MN. Retrieved December 5, 2005, from http://www.ala.org/ala/acrl/acrlevents/ farrell05.pdf

24. Flawn Academic Center: Frequently asked questions. (2006, February 24). Retrieved March 8, 2006, from http://www.utexas.edu/computer/fac/faq .html

25. Freeman, G. T. (2005). The library as place: Changes in learning patterns, collections, technology, and use. In Library as place: Rethinking roles, rethinking spaces (pp. 1-9). Washington, DC: Council on Library and Information Resources. Retrieved August 15, 2005, from http://www.clir .org/pubs/reports/pub129/pub129.pdf

26. Gorman, M. (2000). Our enduring values: Librarianship in the 21st century. Chicago: American Library Association.

27. Gratch-Lindauer, B. (2002). Comparing the regional accrediation [sic] standards: Outcomes assessment and other trends. The Journal of Academic Librarianship, 28, 14-25. Retrieved December 14, 2005, from Academic Search Premier database.

28. Gross, D. (2005, August 8). Birth of a salesman. Fortune, 152, 40-41. Retrieved August 4, 2005, from Academic Search Premier database.

29. Guskin, A. E., & Marcy, M. B. (2003, July/August). Dealing with the future now: Principles for creating a vital campus in a climate of restricted resources. Change, 35, 10-21. Retrieved February 13, 2006, from http://www.pfhe.org/docs/ ChangeFinal.pdf

30. Hazen, D. (2000). Twilight of the gods? Bibliographers in the electronic age. Library Trends, 48, 821-841. Retrieved November 25, 2005, from Academic Search Premier database.

31. Higa, M. L., Bunnett, B., Maina, B., Perkins, J., Ramos, T., Thompson, L., et al. (2005). Redesigning a library's organizational structure. College & Research Libraries, 66, 41-58. Retrieved February 19, 2006, from http://www.ala.org/ala/ acrl/acrlpubs/crljournal /backissues2005a/ crljan05/higa.pdf

32. The Horizon Report (2006 ed.). (2006). Austin, TX: New Media Consortium; Boulder, CO: EDUCAUSE Learning Initiative. Retrieved February 3, 2006, from http://www.nmc.org/pdf/ 2006_Horizon_Report.pdf

33. Kyrillidou, M. (2004, June). Introducing "New Measures." Paper presented at the American Library Association Annual Conference, Orlando, FL. Retrieved March 11, 2006, from http://www .libqual.org/documents/admin/ kyrillidou_ALCTS_2004.pdf

34. Lynch, B. P., & Smith, K. R. (2001). The changing nature of work in academic libraries. College & Research Libraries, 62, 407-420. Retrieved February 17, 2006, from

http://www.ala.org/ala/ acrl/acrlpubs/crljournal/ backissues 2001b/ september01/lynch.pdf

35. Mann, T. (2005, August). Google Print vs. onsite collections. American Libraries, 36, 45-46.

36. Marcum, J. W. (2003, May). Visions: The academic library in 2012. D-Lib Magazine, 9. Retrieved February 26, 2006, from http://www.dlib.org/dlib/ may03/marcum/ 05marcum.html

37. Mathews, B. S. (2004, March). Gray literature: Resources for locating unpublished research. College & Research Libraries News, 65. Retrieved March 19, 2006, from http://www.ala.org/ala/acrl/ acrlpubs/crlnews/ backissues 2004/march04/graylit .htm

38. Mitchell, W. J. (2005, April). Libraries, cities, and networks. Keynote address presented at the 2005 Conference of the Association of College & Research Libraries, Minneapolis, MN.

39. Niebuhr, H. R. (1951). Christ and culture. New York: Harper & Row.

40. Richardson, J. V., Jr. (1992). The gospel of scholarship: Pierce Butler and a critique of American librarianship, including a reprint of Pierce Butler's An introduction to library science. Metuchen, NJ: Scarecrow Press.

41. Shill, H. B., & Tonner, S. (2004). Does the building still matter? Usage patterns in new, expanded, and renovated libraries, 1995-2002. College & Research Libraries, 65, 123-150. Retrieved February 19, 2006, from http://www.ala.org/ala/ acrl/acrlpubs/crljournal/ crl2004 /crlmarch04/shill .pdf

42. Smith, G. A. (2002a). A philosophy of Christian librarianship. In G. A. Smith (Ed.), Christian librarianship: Essays on the integration of faith and profession (pp. 69-84). Jefferson, NC: McFarland. (Reprinted from The Christian Librarian [U.S.A.], 43, 2000, 46-51, 58-59)

43. Smith, G. A. (2002b). A rationale for integrating Christian faith and librarianship. In G. A. Smith (Ed.), Christian librarianship: Essays on the integration of faith and profession (pp. 11-27). Jefferson, NC: McFarland.

44. Stewart, D. R. (2006). Google® of no return? Theology Today, 62, 528-532.

45. Stinson, B. J. (2006). Politically correct science: Why Johnny can't read scientific creationism. The Christian Librarian [U.S.A.], 49, 9-14.

46. Teachout, T. (2006, January 21). A hundred books in your pocket. The Wall Street Journal, p. P14

47. Waller, R. M. (1977). Philosophy and the librarian. N.p.: College of Librarianship Wales.

48. Wand, P. A. (2005). The academic library in 2010: A vision: Report of Symposium 2010. Retrieved January 19, 2006, from http://www.library .american.edu /Symposium_2010.pdf

49. Weintraub, I. (n.d.). The role of grey literature in the sciences. Retrieved March 19, 2006, from http:// library.brooklyn.cuny.edu/access/greyliter.htm

10

CHAPTER

Internet Memes As an Information Dissemination Tool For Libraries: The Ateneo De Manila University Experience

Karryl Kim A Sagun

Ateneo de Manila University, Katipunan Avenue, Loyola Heights, Quezon City 1108 Philippines

ABSTRACT

An increasing popularity of Internet memes can be observed particularly among college students. Websites such as 9gag, Reddit, and 4chan provide a venue for these memes to be visited by the general public. Having observed this trend, the Rizal Library of the Ateneo de Manila University embarked on a strategy to use Internet memes as an information dissemination tool. The initial phase involved creating social media accounts for the Rizal Library to serve as the initial source of information. After which, various marketing strategies were implemented to amplify the followership of the aforementioned accounts, therefore increasing the figures of the targeted audience. The final phase involved creation of the memes and the evaluation of results. After the completion of the project, statistics reflected an exponential growth in the reach of the advertisements posted by the Rizal Library, with the audience themselves contributing to the information dissemination process. This can be supported by the enhanced usage of the services and collections which were promoted using memes.

Keywords: Internet memes; information dissemination; libraries; social media; Ateneo de Manila University; Rizal Library; Philippines

1. BACKGROUND

The Rizal Library of the Ateneo de Manila University kept itself occupied in the past years by continuously building its collection and improving its services. It prides itself in its unique special collections, namely, the Ateneo Library of Women's Writings (ALIWW), a collection of manuscripts and other realia by and about influential Filipino women, the American Historical Collection (AHC), with materials from the American regime in the Philippines, and the Pardo de Tavera Library and Archives, which houses donated materials from influential Filipino families, and is considered a rich collection of materials on Philippine culture. During the past half decade, it has managed to construct two buildings for its expanding collection and growing number of users. Aside from that, it keeps itself abreast with current library practices, such as e-collections, ubiquitous access for library materials, and various digitization projects for preservation and access purposes. However, despite these efforts, statistics reflect low usage for library materials. Announcements also have a very small reach, with very few students knowing about training schedules, changes in hours, and events in the library. Management decided it was high time to give more importance to marketing. Aside from exploiting the gains from the various expenses the library has incurred to service its clients, it is also important to conduct marketing for institutions such as the Rizal Library to gain political, social, and economic support. (Naikwadi

& Chaskar quoting Kotler, 2012). The move was not only necessary; it was also long overdue. The institution then hired an Assistant to the Director for Special Programs and Events in 2011 to focus solely on marketing and promotions. She started with getting to know the different market segments, and found out that the largest mass of the client base was not being marketed to: the undergraduate students. This group was found to be composed of millennials, born between years 1980 to 2000, the era of the computer, and are proficient in working with technology. (Beekman, 2011) They are further described by Solheim (2012) as individuals who "grew up with video games and tend to interact over their smart phones." In the same article it was said that social media plays a vital role in their lives.

It has been evident in the social media sphere that Internet memes have been gaining popularity during the past few years, with websites such as 9gag, Reddit, and 4chan providing venues for these memes to propagate. Masterpasqua (2010) defined memes as "bits of information that replicate themselves; they can exist in brains as well as in any number of other storage devices, from books to the Internet." Thus, it is natural or innate for a meme to replicate itself, even those found on the Internet. This has been the premise of this study: that memes can prove to be an important information dissemination tool for libraries—particularly for promotion of its collection and services as well as for circulating important announcements. Furthermore, Bell (2013) described the importance of using memes in her column, Belltones. According to her, "students of all ages are in tune with popular culture and love tie-ins that show teachers, librarians, and other educators are as well." She adds that memes appeal to people across different age groups. This information served as the basis for the marketing plan of attack to use social media and Internet memes as an information dissemination tool in the Rizal

Library. The strategy was implemented by the second semester of academic year 2011-2012.

2. MARKETING VIA INTERNET MEMES

2.1 Preparation of Social Media

While the Rizal Library did have its own Facebook and Twitter accounts since 2010, the use of these accounts was not very utilized. The Facebook account had only 2,000 likes, while no more than 900 users were following its Twitter account. This is a far cry from the University's approximately 10,000 students. Moreover, the accounts were being used to post traditional announcements, with a formal status update informing the community of changes in schedule, public holidays, among others.

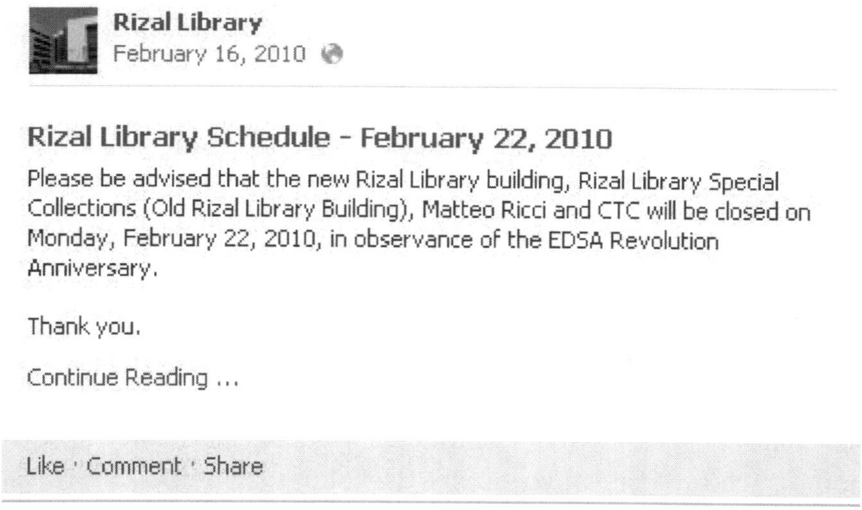

Figure1. Sample of social media announcements before strategy implementation

To improve its utilization, various marketing activities were conducted in 2011 to increase the followership of these accounts. Updating the accounts regularly was the first step conducted. Posting quotes from books, asking engaging questions to help initiate interaction, and holding contests from time to time were performed by the Library.

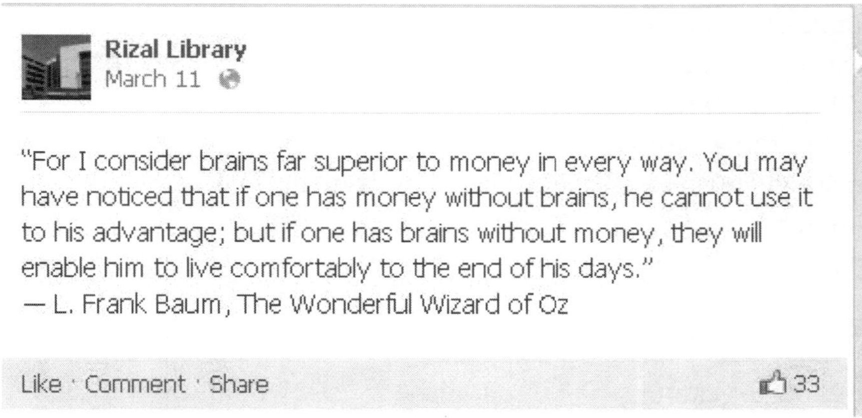

Figure 2. Sample of social media announcements beginning strategy implementation

Followership doubled at this time, and it was evident that the market was ripe for picking—there were enough people to see and share the memes as they are posted.

2.2 Implementation and Results

The strategy initially targeted the most resource-intensive yet underutilized services of the Rizal Library: 1) EBSCO off-campus online database access, and 2) Readers Advisory Service. Being an institution from a third world country, subscribing to EBSCOhost databases and technology services entailed a huge chunk of the institution's budget. It was still, however, found highly important by the Library as it is the "the most-used, premium online information resources for tens of thousands of institutions worldwide, representing millions of end-users." ("EBSCO Publishing", 2011) Readers Advisory Service, on the other hand, was very expensive in terms of labor, as three professional reference librarians were hired to conduct this service.

The strategy employed to use Internet memes was further utilized to market openings for student assistantships, changes in schedule, and new collaborative spaces within the library. In 2011, posters of these memes were posted on the social networking sites of the institution. These were augmented by printed posters situated inside strategic places within the building, such as elevator waiting areas, computer terminals, and doorways.

2.2.1 Marketing Underutilized Services

EBSCO off-campus online database access would allow students to view and download e-articles in the comforts of their own home. Thus, the library is able to provide service even after the working hours of its employees. While this is a useful service for students, there is still room for improvement in terms of usage statistics. Pre-implementation, there were only 1,417 password requests. This reflects low usage if these figures will be compared to the 10,000 students of the University. In the same manner, Readers Advisory Services, or one-on-one consultations with librarians is one of the standard reference services offered by academic libraries, yet the Rizal Library only garnered two during the year 2010-2011.

Table 1. Usage of Underutilized Services

Service	Pre-implementation	Post-implementation	Increase
EBSCO Off-campus Password Requests	1417	1972	39%
Readers Advisory Requestors	2	52	2500%

After implementation of the project, a significant improvement in usage of the two services was observed, as password requests increased by 39%. On the other hand, the number of Readers Advisory requestors recorded a 2500% increase, rising to 52 sign-ups in 2011-2012, after the strategy has been implemented.

Table 2. Extent of Reach of Memes for Underutilized Services

Service	Likes	Shares
EBSCO Off-campus Access	406	285
Readers Advisory Service	545	147

The memes for the two aforementioned services were posted on the Rizal Library's social networking accounts. The poster for the EBSCO off-campus access garnered 406 "likes" and 285 "shares" on Facebook, while the poster for Readers Advisory Service acquired 545 "likes" and 147 "shares".

2.2.1 Use of Memes for Announcements

The Rizal Library has had openings for student assistants for the past couple of years, but had no applicants for the posts. In the same manner, students were being informed by the library of changes in schedule, but only a few have been reached by the said announcements. Thus, it has been decided that the same strategy implemented to promote the aforementioned underutilized services be used for important announcements as well. Results show a positive outcome, as post-implementation, the institution has received 91 applicants and counting.

Table 3. Extent of Reach of Memes for Announcements

Service	Likes	Shares
Term Break Opening	317	71
Roodeck Study Hall Availability	119	105
Student Assistantship Program	251	75

In applying memes to its announcements, the Rizal Library has achieved a farther-reaching effect for its posts, with its audience "sharing" or "reblogging" the posts, hereby helping extend the reach of the announcements. Announcements have also increased their reach exponentially, with 317 "likes" and 71 "shares" for the term break schedule announcement, and 119 "likes" and 105 "shares" for the availability of the Roofdeck Study Hall.

3. CONCLUSION

Memes proved to be a very effective means of conveying a message to the Rizal Library's clients. With an expanding followership for its social media accounts, and a continuing popularity for Internet memes, the Rizal Library intends to continue using this medium to promote its other services and collection. In the same manner, the institution shall continue to use memes to post announcements and exploit the multiplier effect of this method. The Facebook account of the Rizal Library now has over 7,000 likes, and its Twitter account has garnered over 5,000 followers, and an increasing trend is foreseen for the academic years to come.

REFERENCES

1. Beekman, T. (2011). Fill in the Generation Gap. Strategic Finance, 93(3), 15-17.

2. Bell, M. (2013). What I Really Meme Is .. Internet@Schools, 20(2), 24-25.

3. EBSCO Publishing. (2013) Retreived May 8, 2013, from http://www.ebscohost.com/

4. Masterpasqua, F. (2010). What are memes, and what do they contribute to our understanding?. Psyccritiques, 55(47)

5. Naikwadi, V. A., & Chaskar, P. M. (2012). Implication of marketing plan: for marketing library and information services. Indian Streams Research Journal, 2(10), 1-6.

6. Solheim, N. (2012). MIND THE GAP. Benefits Selling, 33-37.

11
CHAPTER

Access and the Changing Mission of Research Libraries

Hans-Joachim Wätjen

Oldenburg University Library P.O. Box 2541 26015 Oldenburg, Germany

FROM BABEL TO VIRTUAL LIBRARIES

The session of this morning deals with the coming century. But I am not a prophet foreseeing the future. I agree with the economist Edgar Fiedler who said in his Law of Prediction, "Forecasting is very difficult, especially if it's about the future." And Mirjam Foot is on the right side too. In the afternoon she will talk about change as the only certainty. Or in other words: today we live in an age of uncertainty and it will increase in the future! Before speaking about the future of libraries I would like to go back to the past.

Ever since the fire of the Alexandrian Library scholars have been dreaming of the all-comprising collection of books. Feudal lords, scientists, librarians, poets and ordinary people have wanted access to the complete knowledge of mankind. But the complete and universal library was never again meant to have its place in the real world. It was only meant to be there as a dream, such as in the phantastic fiction of Jorge Luis Borges and his Library of Babel. I quote:

> „When it was proclaimed that the Library contained all books, the first impression was one of extravagant happiness. All men felt themselves to be the masters of an intact and secret treasure. There was no personal or world problem whose eloquent solution did not exist in some hexagon." [1]

Nowadays, we would say: on any kind of Web server. The hope for this particular feeling of happiness, curiosity for the unknown, but also the intentional search for important information are the underlying forces for millions of people all over the world to travel virtually in the World Wide Web. Yet, the all-comprising virtual library will not be available in the near or far future, not even in our modern society of knowledge and information. There are far too many printed books in our stacks, and new ones are being added daily. Every year the book-fair in Frankfurt offers more new titles and the publishing industry is still growing by selling printed books, journals and newspapers.

The traditional book offers an almost perfect interface for users - almost perfect, because I can very well imagine a foldable waterproof plasma screen for reading in the bathtub. But would I like to take into my bed the screen and stand-by batteries as well? No - Marshall McLuhan's prognosis for the death of the book was wrong and will remain so! And, as a consequence, the following cartoon by Til Mette2 will not turn into reality, either:

Figuure. 1

The future of libraries and librarianship has been discussed controversially for some years and the debate is still going on. On the one side theorists of our profession together with some scientists, science administrators and many library users believe in the progress of information technology to solve all the problems of information provision. Access is the code word for their virtual library as expressed by the motto of the 6th WWW conference 1997:

"Accessibility ... Everyone Everything Connected". It seems to be so simple: just clicking links to get information.

Within this group the optimistic colleagues believe that the library can be transformed to an information warehouse and Web portal as a one stop shopping site for information. Their library will be run by library managers, IT professionals and modern librarians teaching information literacy during the transition period until users have turned to end-users. The pessimistic colleagues within this first group foresee the death of the book, of the library and of the profession. Their world of information is totally virtual and dominated by Microsoft, AltaVista, Elsevier, EBSCO, Bertelsmann and other major players.

The arguments of both groups are helpful to analyse what is really going on. They are wrong and right. Between these groups the more practical librarians know that the library of the future has to find a cost-effective balance of access and holdings. Their library will be a hybrid one integrating real and virtual collections. They fight for higher budgets to invest in both: in real collections of locally stored paper and digital resources as well as in IT and services for better access to virtual collections. But more and more they feel bad about fighting for shareholders of Elsevier and of other publishing houses.

For both groups it is clear that the methods and techniques of communicating, producing, distributing and imparting knowledge now change rapidly and radically. All of us know that we have to redefine the traditional role of the library: what and how to select, to acquire, to classify, to catalogue, to provide, to archive or to give access to and how to assist people in the use of information and more important: how to provide free and equal access to information according to the mission of libraries.

Showing the present reality of virtual libraries I will further on talk about the future trends of information media within higher education and research to discuss possible tasks and consequences for libraries at the beginning of the next century

VIRTUAL, DIGITAL, ELECTRONIC - WHAT LIBRARY IS IT TO BE?

The reality of virtual libraries can best be experienced within the World Wide Web itself. Searching for the virtual library there you will find a touristic advertisement for the Truse Valley under the most promising address: .

Figure. 2

Today's clever marketing people, hoping for as many clicks as possible, have simply bought the right for this domain name.

Further and more thorough searching will inevitably lead to another dilemma. With the help of the German search engine FIREBALL you will find 3,300 hits for "virtuelle Bibliothek", 1,300 hits for "digitale Bibliothek" and 600 hits for "elektronische Bibliothek" - all this even without truncation and only in Germany. With the use of AltaVista and extended by the English phrases the result will be more than 270,000 entries.

With this number of hits in mind, we may well conclude that the use of language by librarians within the World Wide Web corresponds to that by any other netizen: all three terms are being used arbitrarily for the whole variety of information offered on the Internet. The net reality shows that librarians, normally used to working with controlled vocabulary are not being helpful in defining the differences between an electronic library from a digital or a virtual one.

On the other hand, our colleague Mr Hilberer from Düsseldorf is making an attempt at a suitable definition. I quote:

> "Digital libraries are collections of electronic information in the possession and thus under control of real libraries. Virtual libraries are collections of references (collections of links) to information which as such are not in the possession of the corresponding real library. By their nature, all the virtual libraries are digital, too, but by no means are digital libraries virtual in all the cases. For instance, collections of CDROMS are non-virtual digital libraries." 3

According to Hilberer an electronic thesis from Düsseldorf would be part of the digital library in Düsseldorf, but part of the virtual library in Oldenburg. The CD-ROM "PsycLit" would be part of the digital library, but access to the very same data via FirstSearch of OCLC would be part of the virtual library. Technical facilities do away with borderlines as well: an analog video e.g. only to be played electronically, turns into digital and virtual information "on the fly" when attached to a Web-TV-server. And even the real book borrowed from the University Library of Düsseldorf, can turn virtual if it is not returned, in spite of reminders. We should not confuse users by giving confusing names to information. Distinctions by technology, cost or source are not helpful. The user does not want to know whether an information resource is part of the electronic, digital, virtual or real library. He wants to access it for searching, reading, copying or whatever.

The meaning of "virtual" as defined by Microsoft's Encarta comes exactly to the point:

> ‚virtual'
>
> 1 available as a possibility
>
> 2 as a thought, appearing to be4

The virtual library is in existence as a possibility, with one or two links functioning or not functioning, with one information not costing anything, with another causing costs, with the document delivery supply unreadable, but with the possibility that it might have been readable. The second meaning is true as well: the virtual library is in existence as a thought, only appearing to be in existence. Because, in spite of hundreds of millions of Web pages, today it is only a tiny library, compared to our real libraries. It seems that, in view of the 270,000 hits, and even after separating the number of duplicate available, we have more virtual libraries than virtual books in the net. Or, put differently: we have too many virtual libraries and too few virtual books. And, at least partly, librarians are to blame for this situation. There is hardly a library that does not offer its local CD-ROM collection, its more or less well commented, incomplete, manually composed bookmark lists, or its Web interface for their catalogue under one of the three types of language on a world-wide level. We cannot accuse the local pizza baker of not having the meta-tag "noindex" in his Web pages, or of his server directory not including a "robots.txt", a command for the robots of the search engines.5 But the information experts of the libraries ought to know that their Web list of only locally accessible data increases the recall of AltaVista and many other search engines to such an extent that it obstructs the search for the original supplier of PsycLit, Inspec and other data bases.

The situation looks definitely more sad if we have a look at what libraries are not offering. The great number of grey literature by scientists working at their own institution is in deep slumber on the hard disk of their personal computer, on diskettes in the drawers of their offices, or, at best, on the Web servers of institutes and faculties, and unnoticed by librarians. You cannot gain access to them by the online catalogue of the library or by any of the regional online catalogues. The needles in the haystack are either lying on the doorstep, as yet ungathered, or they have already been gathered by information amateurs like former Stanford students Jerry Yang and David Filo, the makers of Yahoo! and today's billionaires or by scientists like Ginsparg, the founder of the physics pre-print servers.

The same is true for printed treasures in our stacks, whose copyrights have expired. We might as well recover them and put them on the net instead of paying collectors' prices for facsimile editions. But hardly any library is willing to do this without project sponsoring. Again, it had to take amateurs like Michael Hart, Gunter Hille and many voluntary non-librarians to found the Gutenberg-Project6 and clever publishers like those from

Directmedia Publishing, who are marketing the "Digital Library with German Literature from Lessing to Kafka" on CD-ROM and on the Internet7 , and who have snatched away the domain name "digitale-bibliothek.de" from German libraries.

It seems that librarians are always on the spot a little too late, but always with professional thoroughness and solidity, hyperactive in many projects and driven by their apprehension that the status of their profession and the library as an institution could be shoved aside by others in the World Wide Web.

If we look at the results of librarians and publishers building the virtual library, both have not worked very effectively if we compare it with those of the major players in the Internet. There are only few exceptions like some national academic Web indexes, subject based information gateways and digitised collections which were created and well maintained by librarians. I hope that we get - or better take - a second chance for the next steps on the way to virtual libraries. Before we look at the concrete chances and the possible new mission of research libraries I will give you some of my prospects for the technological change and the future of the information media.

FUTURE TRENDS IN IT

It is not the time to discuss the future trends intensively. I will give you only a short list of my forecast in the field of IT:

We will live more rapid technological change:

- Today the capacity of storage media is no longer a real problem if we think of DVD, personal computers with huge hard disks and servers with terabyte archives. The problems of limits in the past will be unknown in the future.

- The bandwidth of our local and wide area networks will be great enough e.g. to transport streaming videos in high resolution full screen quality and three-dimensional virtual realities.
- The mobile hand-held communication will be integrated with mobile computing. We will take it as a matter of course when students will visit the virtual library from everywhere e.g. from their tables in front of the coffee bars.
- Software will be more user-friendly and more suitable driven by voice commands and voice outputs. Helping agents and robots will assist users personally. But the speed of change will bring new bugs with new software applications as well as it does in the present.

FROM BOOKS TO MULTIMEDIA PRODUCTS?

Let us stay with the future of information media for a few more minutes. The complete text of Heine's Wintertale on my PSION hand-held computer is nothing but a gadget, like the "Rocket Book" or the "Soft Book" as new devices for electronic books. Today they are toys for big children. The English dictionary which I have stored on my PSION, on the other hand, is rather useful for consultation. Electronic publications have already replaced their printed counterpart in the field of reference works and boring bibliographic indexes, and they will soon replace scientific articles in journals and similar sources of information.

Neither librarians nor users mourn the loss of the numerous volumes of bibliographic reference works - only, if at all, our administrators of CD-ROMnets do. Almost daily, they are angry about badly programmed and incompatible software, not to mention their trouble with Microsoft's operating systems.

With the increasing number of software and multimedia products in the field of teaching and learning, the number of technical problems will increase. Looking at our kid's computer games with their particular hardware demands and destroyed operating systems after installing the new game, we get a glimpse of what the ideal world of multimedia - so much yearned for by educational politicians as well - might be like.

Librarians who have considerably neglected or only reluctantly touched the predecessors of this type of information, the video films, will no longer be able to keep away from these sources of information as they will replace existing textbook collections in many fields. The consequences for the libraries will be the more serious the greater the span of time before we start dealing with them and being able to handle them competently. Otherwise, this job will be done by others, by the data processing centres and by the media centres. Or else, the collections will again be under the administration of the individual institute's secretary or of the faculty's science assistant - nicely decentralised on a CD-ROM or on the Internet, but by no means retrievable in any library catalogue.

Or the change will be much more radical. A scenario of academic education in the future just published for the Bertelsmann and the Nixdorf Foundations gives an impression of the market for higher education in the year 20058 : The student's decision will not be to study in Munich because of his girlfriend or in Oldenburg because of the lower costs of living. He will search the Internet for online suppliers of higher education globally and he will find offers of certified distance-learning modules, course-packs and "training in the box" by international consortias, corporate universities, virtual universities and networks of universities. The study says that 50 percent of the students will not be educated at traditional universities anymore. Their study will be Webbased, flexible, configurable, interactive with simulations and computer supported co-operative work (CSCW). The students decision will be costbenefit oriented because he will have to pay for his education 15,000 Euro annually. In this scenario of competition between universities and other suppliers in the education market the authors do not mention the library whether playing a role as a distributor and technical provider or not. Today we see millions of federal money invested in multimedia developments at our universities. I am not sure, however, if the prognosis of Bertelsmann and Nixdorf will succeed either in its business idea or in the complex reality of learning. I hope that the more behavioristic approaches of learning and teaching by multimedia will fail and that the idea of constructivistic learning with its social dimensions will succeed in the sense of Benjamin Franklin: "Tell me and I forget, teach me and I remember, involve me and I learn." Multimedia and Web based learning will come but it should be used as a tool for better communication and information to reach more creativity and involvement of the students and a better quality in teaching. The potential of multimedia will also reach the research publications which are still electronic copies of the linear paper versions.

Considering that today the number of electronic editions of monographs has so far reached a level of under one per cent, and that the number of journals is far under 10 per cent the current status of the virtual library is indeed a relative one.

My prognosis for the immediate future is different. Textbooks for teaching will be supplemented and partly replaced by multimedia products. Typical research monographs for a small but world-wide circle of readers, scientific journals and serials, bibliographic and other reference works will primarily be offered electronically. Only one question remains: by whom and at what financial and legal conditions or in other words:

FROM PAPER TO PAY PER ... ?

We already have heavy turbulences and conflicts among authors, publishers, booksellers, retailers, libraries and readers. During this transition period all actors in the information chain try to use a double strategy: they conserve their traditional roles and they call each other into question.

Here I cannot discuss the serial crisis very deeply which is illustrated by the following cartoons9 :

Figure. 3

Figure. 4

The library's mission to offer access at non-prohibitive prices for everyone and for a long time differs from the shareholders' and managers' interests of large publishing firms. They have started to create an e-commerce library. During the transition period they bundle print subscriptions together with online access at additional prices. They try to increase their share of the market by selling access to all titles if the library accepts the "no cancellation" clause. All these previous price policies of the publishers will escalate the crisis. Maybe the escalation is necessary to bring libraries and the scientific community to another policy rethinking their purchasing power and their property rights

But today the reactions of libraries and scientists are inconsistent. On the one hand the libraries have formulated statements and guidelines for the selection and purchase of electronic information like those of the "International Coalition of Library Consortia" 10 with 5,000 member libraries world-wide or those of Dutch and German university libraries11 or the new licensing principles of LIBER12. On the other hand they ask the state for extra money and if they get it they sign the contracts with Elsevier and Academic Press under terms and conditions which do not accord with any of their principles. It is the same with the scientists. They still give their articles to the commercial publishers, establish new journals for them, and do not think of the consequences or about alternatives.

Instead of doing this we should act together with the scientists and the scholarly societies. We should develop and practice new models of scholarly publishing using the possibilities of the World Wide Web and of new technology. The attempts of Stanford's "High Wire Press" or the alternative concept of the "New Journal of Physics" or "SPARC - The Scholarly Publishing & Academic Resources Coalition" 13 lead to the right direction. They try to create a more competitive market place for research information by publishing new journals within scholarly institutions, by founding a nonprofit publishing house or by establishing self organised structures for electronic publishing.

A scenario a little bit more radical would be the following one: What would happen if all the members of LIBER together with all the 5,000 members of the International Coalition of Library Consortia decided not to renew the subscriptions of the five main publishers with the highest price increases for the next year and the learned societies agree with that boycott. For the meantime the planned articles could be refereed as usual and then send electronically together with their meta data to a central database maintained by SPARC or to decentralised databases at the national libraries or to the library of the authors' home institution or to all three.

I fear that this scenario will remain a dream. Libraries still hold subscriptions even of the nearly criminal publisher MCB

Publications and librarians accept the same publisher still as a main sponsor of IFLA which can finance round tables about the future of electronic journals as it happened.

But let us now look forward a little bit more optimistic at the reality and at the concepts for the future of libraries which will be hybrid ones. In these times of transition, the question for libraries arises as to how the new information media will have to be presented to our clients together with the old and still new printed information.

THE HYBRID LIBRARY: INTEGRATION OF INFORMATION AND WORK

"I agree with the British colleagues who have formulated their idea of a hybrid library as follows: "Communications and Information Technology enables the integration of information and work." 14 Another description comes from the Santa Fe Workshop on Distributed Knowledge Work Environments: "[T]he concept of a "digital library" is not merely equivalent to a digitized collection with information management tools. It is rather an environment to bring together collections, services, and people in support of the full life cycle of creation, dissemination, use, and preservation of data, information, and knowledge."

Students and scientists want to do all the things in the virtual library that they have always done in the real library - but with digital means, without any frictions and more comfortably, without having to leave their seats and even working from their homes. They want to find, read, hear, see, store, print, analyse and adapt information. They want to criticise and comment on it, to quote it and to take it into their individual database. They want to send it on to colleagues and if necessary they want to pay for precious information and services.

Each of these activities is connected to a number of individual processes. Just think of the searching and finding of information in the Internet. The problem is not a new one. I quote Borges again:

As was natural, this inordinate hope was followed by an excessive depression. The certitude that some shelf in some hexagon held precious books and that these precious books were inaccessible, seemed almost intolerable." 15

Borges' idea of the Universe as an infinite library with maze-like chaos is comparable with today's reality of the Internet. In spite of the great number of search engines, we have great problems sifting the chaff from the wheat, or to find the needles in the many haystacks. Many steps are necessary for finding the information as such. Just think of your own library, of what searching, what change of systems and, at best, what fussy copy and paste activities are necessary. As a rule, the user has to write down new references over and over again in order to insert them in their next search. If they choose an interdisciplinary subject, they are even worse off.

And up to now, we have only dealt with the finding of information. If the user wants to produce new information for the virtual library, the situation will be even more complex, such as in this example:

"Supposing a scientist open to modern techniques would like to put into practice the appeal of the Heads of Universities for the integration of new media into university teaching. He would like to produce a new teaching and learning unit in his particular subject which is to be based on multimedia, in the process of which he has to prepare his own texts, printed texts, graphic representations from books, taped documents and video material as well as bibliographic references from a didactic point of view, and he has to present them to his own students and to those of other universities on an interactive hypermedially connected level in the World Wide Web.

Full of hope, he asks for technical support at the data processing centre, the library and the media centre. He would then be handed on to various people and - if he was lucky, get to know half a dozen helpful colleagues in different institutions. At best, he would be given hints at how to procure financial means from national and European sponsoring programs for his institute. But it is not very likely that any of the universities would be able to fulfill the services he would be in need of." 16

It is part of the hybrid library to offer services for these requests. If the library is not able to do this on its own it should outline an integrated service together with the data processing centre and the media centre and introduce this service in a clear and understandable way to the user.

My virtual dream library consists of an integrated access to any kind of information as well as of the workplace for integrated use and processing of information - in the library, at the institute and at home. This vision is far from being a reality. It is not enough to put up 200 personal computers in the library, to equip them with any kind of software and to list the sources of information via the Web browser. Colleagues from Groningen University Library have made the experience, that this only leads to surfing on the World

Wide Web or to using the personal computer as a digital type-writer. Students are not to blame for this. It is also due to a lack of integration of these precious sources of information that they are often not being used. Therefore, let us have a closer look at the information on offer.

INFORMATION IN THE HYBRID LIBRARY

In the hybrid library the WWW is to serve as a mean to make various sources of information accessible, independent of place and time and of their physical form - in particular:

- the own OPAC and those of other libraries,
- the shared regional catalogue,
- the national catalogue, if there is one, and those of other countries,
- data bases,
- printed and electronic journals,
- printed and electronic books,
- multimedia material,
- special collections (maps, audio, video, photo archives ...),
- retrospectively digitised material of the own library and of others,
- other sources of information in the World Wide Web,
- collections of data, facts and software.

And all this irrespective of the resources being available in your own institution, by remote access via the Internet, or by interlibrary loan and document delivery services.

Many of these resources are already being offered by libraries via their Web interfaces. But the following polemic remark by Umberto Eco, originally meant for traditional libraries, is also true for hybrid libraries:

"Catalogues must be separated as far as possible: we should be particularly careful to separate the catalogue of books from that of journals and that of the journals from the subject catalogue, the catalogue of new acquisitions from that of the older collection." 17

As a rule, there are still different types of access, of systems and of user interfaces, which are offered by libraries, depending on the catalogues, on the data bases, on the publishers, etc. They are often to be found in seperated workplaces, too. There have been a variety of technical and organisational reasons for this phenomenon. For the librarian at the information desk and even more so for the user, the services available are hardly surveyable or conveyable. It will not do to make a list of the possibilities of access to different information resources and to their respective WWW gateways on a single HTML page. Many suppliers of information and many libraries as well have taken this road, but without achieving a true integration of different sources of information. The solution cannot lie with one interface and with one system, as the sources are too heterogeneous. Our aim should be to create few integrated Web-based accesses of selected high-quality resources. Let us have a look at the individual possibilities and let us start with the library catalogue.

FROM THE OPAC TO AN INFORMATION GATEWAY

Today the local catalogue can be accessed via a Web gateway. But navigation possibilities via hyperlinks are simply being given away in many solutions with Web OPACs. The hits they indicate contain HTML only as "preformatted text" without any links to connected titles or index entries in the data base. But it could be made possible to refer from one indicated hit to others of the same author, the title key-word, the subject index, the classification by means of a link. Even corporate entries, hardly ever searched for by anyone, could be made useful by a click. Most of these information are available in the local data base - they would just have to be supplied as a link via the perl scripts in order to release a new search script on click.

But most of all, the Web OPACs offer a decidedly new quality by making it possible to gain access to external information by means of a link: direct access to full text, to the table of contents of a journal, to the abstracts of a CD-ROM data base, to the publisher's server with its electronic version of the journal, but also to pictures, audio data and videos. Many of these sources of information are digitally available in the library or accessible on a licence basis, but not yet connected to the OPAC. The local catalogue could thus turn into an information gateway to the hybrid library. There have been examples, though incomplete, for this kind of development: The University Library of Bielefeld links the search in the catalogue with a search for data on articles (JADE), on abstracts (Silver Platter), on full texts of publishers, on their own etrospectively digitised books, on Web sources (IKAT), and with the document-supply service Jason:18

Everyday the OPAC of a library is being used thousands of times by library users whereas expensive CD-ROM databases are being left relatively unnoticed. Integration into the Web OPAC can bring about a more intense use of the data. Technical solutions exist, such as ERL by Silver Platter and a gateway for Ovid databases. Both of them make the links between reference information and concrete holdings possible.

Dokumente: 1 - 20 von 31

Nr	Datenbank	Titel
1	IKAT	Social Science Information Gateway
2	OPAC	Social sciences in transition
3	OPAC	Evaluating information
4	OPAC	Modes of perceiving and processing information
5	OPAC	Directory of social science information courses
6	OPAC	Selective inventory of social science information and
7	OPAC	World social science information directories
8	OPAC	Directory of special libraries and information centers
9	OPAC	Social science information
10	OPAC	Social science information and public policy making
11	OPAC	Evaluating information
12	OPAC	Latin America
13	OPAC	Political science, government, and public policy series
14	JADE	Social Science Research Needs and New Information Technologies
15	JADE	Global Society and Information Technology: Social Science

Figure. 5

Libraries which have a licence agreement with publishers for electronic journals usually find that the possibility of direct access to articles is being used too little. Normally, these libraries offer a list of the electronic titles in the World Wide Web, in alphabetical order and according to subjects and publishers. Even if we establish a link in the OPAC between the titles and the publisher's server - as we have done in Oldenburg19 - the demand for this service is very limited.

If, on the other hand, the bibliographic data of each article in the OPAC were to be mixed with links to abstracts and full texts, demand for this service would surely increase considerably. But hardly a publisher would be ready to supply one single library with data on a regular basis.

The alternative would either be for a large consortium to feed the data into the bibliographic utility system at the regional or national level or else for us to go and get us the data ourselves from the publisher's server by means of a robot such as Harvest. This second path is being taken by the University Library of Osnabrück in its "Electronic Library" project.20 It is thus collecting the accessible pages of publishers' journals and the data of freely available electronic journals in the Net, a few thousand titles by now. During this process of integration into the OPAC data base, the problem of consistency of the data has to be solved, not a trivial one, even with programs for checking links. There is also the need for regular checking that the robot is not gathering completely different information, because the structure of the publisher's server may have changed in between. The same applies to inserting any other Web resources into the OPAC, whether manually or gathered by a robot. I would rather warn you of it. There is a different solution, instead. By means of a gateway search enquiries can be sent simultaneously to the WWW OPAC and to a search machine or a Web index. This can be implemented without great expenditure. The great success of meta crawlers, but also of the Karlsruhe Virtual Catalogue21 lies in the simple principle of simultaneous enquiry by scripts.

Not all sources of information can be integrated into this model of simultaneous enquiry. The number of potential services has become too big for this. "Internet sleuth" 22 alone offers gateways to over 3,000 data bases in its own data base, and has thus turned into the data base of data bases in the Internet. I warmly recommend the use of this service to any librarian at any information desk. The link to this service should not be missing on any home page of any hybrid library.

CATALOGUING RULES AND DUBLIN CORE

When processing electronic documents and, in particular, Web resources we should examine the cost factor carefully. We can simplify the work if we let the authors at our institutions do the simple cataloguing. The Dublin Core Metadata Element Set has gained considerable ground internationally with such a process although its use is not widespread. There are already too many templates for the recording of the Dublin Core Data in the Internet. You should not develop your own method, but rather use an established one. Whether you simply convert these data into the national MARC or into your own format and transfer them unchanged into the OPAC or whether you want to check and complete them is a question of feasible cost-expenditure.

A problem is presented by the rules for cataloguing. They prescribe that for each new edition of a publication an individual record of title must be made. The result is then quite curious. If you imagine that, because of the deterioration of the paper, you film an old book from your library which already has a title record. Then you digitise the microfilm which you have produced and deposit it on your document server as an electronic publication. The result is then four different records of title: one for the original, one for the master film, one for the user film and one for the electronic document. In your Web OPAC the non-circulating original will probably be shown first and in final position the electronic version, which can be accessed with a click, will appear - provided, of course, that the user will scroll down that far. I think that in this case a pragmatic revision of the existing rules is urgently required

FROM INFORMATION TO SERVICES

In the hybrid library the capabilities of the WWW are still too little used for providing services. Many libraries offer services via interactive Web sites which can be introduced at one's home library without great expense.

Requests for information can be answered not only by telephone or by walking around the library, but also by sending a Web-form immediately and automatically to the reference librarian or another responsible staff member per e-mail. Do not be frightened that this service will be used very often. The users' goal is not to get answers one hour or one day later but to get on with their work. Today the Internet-using clients are answering more questions on their own. We have to do the next steps "providing a point-of-need "reference service to information seekers at the place where they are when they have a question." 23 The reference service must change in the hybrid library to an interactive remote service even if the user is working in the library but fears to leave his workstation and to loose his chair when going to the help desk. There are possible techniques to provide a just-in-time reference service. A flashing link on all our Web pages including the dynamic ones of our database gateways can offer a talk to a librarian by different ways: by talking via CuSeeMe or Netmeeting, by chatting via a second browser window. The online interactive help service could be started locally or it could be organised co-operatively by several libraries at a regional or national level to offer a 24 hours service.

There are also library services which intend to replace or to support the work of reference librarians. The university library of Groningen developed a system called COBRA which makes suggestions for search strategies according to the needs of the user.24 It is an expert system which generates an individual advice of all kinds of reference works, books, databases, abstracting and indexing services, by using a local database with more than 3,000 reference titles.

In addition to the concept of integration by simultaneous gateways as described we need a knowledge database or an expert system for our users and our reference staff as a gateway to gateways to fill the lack of information literacy and to mediate the different resources according to the needs.

Another example: The scholar no longer has to send his suggestions for book orders to the library via the slow in-house mailing system. She or he can send the bibliographical details of the order directly to the respective subject librarian via a Web-form.

The almost complete availability not only of the faculty members, but also of many other users via e-mail opens new possibilities for providing information. The New Books List can be sent to subscribers according to individual subjects without burdening the library with additional work. Current contents information aimed at the needs of individual users can also be distributed in this way. Other automated SDI services can be pushed to the users. Ultimately costs can be saved if the overdue warnings of the circulation department are delivered automatically and without delay via e-mail.

One principal service area does not lie exclusively in our hands, but can be realised only by co-operative effort. I address here the functioning of interlibrary loan, or, as we say in the hybrid library - the document delivery system

DOCUMENT DELIVERY AND SUBITO

In Germany we have a document delivery system that consists of five different access systems. Although I was personally involved in the initial phases of the planning of SUBITO, I must confess that I only recently placed my first order via the SUBITO access system of the DBI.25 The experience was astonishing. Since my information about the journal article was incomplete, I first of all searched for the article data in a Current Contents Database and found it. I copied the data in the clipboard and switched to the access system of the DBI. After the troublesome registration and receipt of a password I was also able to find the journal in the National Serials Database. From a long list of scarcely interpretable data on holdings I selected a large library and laboriously inserted my article data from the clipboard into a form and ultimately sent the order off. The e-mails, which reached me after that, were cryptic and incomprehensible. I interpreted the first as a confirmation of the receipt of the order, while the second announced that the delivery was to be expected forthwith. The third was supposed to be the delivery itself. All e-mails seemed to be "letters" automatically generated by the document delivery station, which contained numerous numerical codes and began without a salutation and ended without regards. The delivery itself contained an attachment, which could not be opened by any of my viewers. A glance into the file with a programming editor showed that it was only a few bytes in size and did not contain the Megabytes of the expected multi-page TIF file. As indicated, I sent my complaint that the TIF- file was probably damaged per e-mail to the address indicated with the request for renewed delivery and since that time I have heard nothing more from SUBITO. I have also never received a bill. The example is, I hope, an individual case, but nevertheless instructive.

A further problem with SUBITO is identification. If we require simple accessing procedures via the domain from publishers and database providers, the librarians themselves cannot introduce a system full of complex log-in procedures and multiple passwords and then give the whole complex tangle a single name.

In my mind a user-friendly service of the hybrid library is something quite different. An easy entry into the regional and national system for ordering and delivering documents should be offered via the local Web OPAC, which should also encompass the holdings of the other local libraries. The user-code of the home library should suffice for identification and the article and journal data should be immediately transferred into the ordering system. The SUBITO services JASON and GBV approach this system in their configuration and structure.

ON DECIPHERING AND CONVERTING: DIGITAL ARCHIVES

Let us think 20 or 100 years ahead. We will have a problem which Jose Luis Borges foresaw nearly 60 years ago:

> "Five hundred years ago, the chief of an upper hexagon came upon a book as confusing as the others, but which had nearly two pages of homogeneous lines. He showed his find to a wandering decoder who told him the lines were written in Portuguese; others said they were Yiddish. Within a century, the language was established: a Samoyedic Lithuanian dialect of Guarani, with classical Arabian inflections."

We would say today, that in the year 2099 the restorer in the workshop of the German Library in Leipzig luckily succeeded in decoding the file format of a Microsoft Winword 8.0 file on the library′s document server with the help of a virtual Windows98-engine and made it readable. Whether that will actually succeed in a hundred years, nobody knows. The long-term filing of digital documents is a pressing problem which the virtual library has not yet solved. The chances of success will be all the greater if various copies in different file formats are preserved and still available on our servers. No one knows whether it will be easier to convert and decipher Postscript Level 2 or PDF Acrobat 4.0 or Winword 8.0 in a hundred years. It follows that if you establish a collection of important electronic publications in your library - for example the publications of your institution - you should never delete the source format after having converted it into another. Today's html-, sgml- and future xml-files will surely be able to be read by a programmer with a manual from the year 1999 who can write an appropriate parser. But what about the other file formats and those which Bill Gates will present us with in the future? No one can safely assume that the National Library will solve all the problems of deciphering files. The problem can only be solved by co-operation among all research libraries.

FROM PROJECT SUPPORT TO THE PLANNING OF THE HYBRID LIBRARY

With the problem of digital archives we have arrived at the end of the daily routine in the hybrid library. In closing, I would like to come to the national library policy. Arnoud de Kemp, a manager of Springer, said at a conference: "There are too many research and development projects for professional societies, publishing houses, libraries and the disseminating book-trade." 27 As regards Germany I would say that there are too many expensive projects whose benefit for our customers is questionable and that these projects do not co-operate, but rather compete with each other. In Germany we lack coordinated and professional planning for the development of the hybrid library.

There are certainly enough studies and papers available for such planning. The Ministry of Education and Research, the German Research Society, and the Research Council have worked out numerous recommendations and formulated programs. Co-ordination and a concrete planning concept must still be achieved in Germany. Federalism cannot be an excuse for this situation. It happens too often that committees meet to discuss these very same questions on the federal level. "Global Info" 28, which now calls itself "The German Digital Library Project" is being funded by the federal state. Simultaneously, the Research Society supports three seperate programs in the field of "Deployed Digital Research Library".

All these initiatives exist side by side. The supported projects themselves do not seem to be subject to any real co-ordination or supervision. It is, for example, incomprehensible why the digitisation of ancient Egyptian dictionaries or the documentation of the television program, Black Channel' during the cold war should be important elements of the "Deployed Digital Research Library". It is obvious that still more co-ordination is necessary - above all the development of pragmatic, practice-oriented concepts according to the user's needs.

The most recent criticism by foreign experts of German scientific development can also be directed at the library development of the ministry and the DFG: It does not suffice to establish programs and then to wait and hope that corresponding grant proposals will be forthcoming. The planning and realisation of the infrastructure necessary for the hybrid library can not be left to chance. Those who are involved in developing the hybrid library must be asked specifically to undertake the necessary projects and they must also be required to act co-operatively and to organise the necessary procedures for professional planning.

Our British colleagues have shown us that this is possible. In phase 3 of the program "e-lib" 29 the concept of the hybrid library is pursued, which has as its principal goal the integration of information resources and the work of the research scholar. While in the USA the "Digital Library Initiative" 30, financed by the National Science Foundation, is strongly oriented toward research and new technology, the program in Great Britain is oriented toward practical application and is therefore at the same time both visionary and practical.

A FINAL REMARK

Let us give the word again to Jorge Luis Borges:

"The impious maintain that nonsense is normal in the Library and that the reasonable (...) is an almost miraculous exception." 31

My lecture was not intended to be irreverent. On the contrary: The criticism of existing incomplete solutions should bring us closer to Borges' paradise, which he imagined as a kind of library, for us perhaps the ideal virtual library which is a hybrid one. I have the impression that librarians shy away from open criticism today. In earlier times they often earned their money with brilliant and biting literary criticism. We should also become critics again - in and through the Internet.

REFERENCES

1. Borges, Jorge Luis: The Library of Babel - Translated by J. E. I.

2. Stern - 19.6.1997

3. Hilberer, Thomas: Was ist eine Virtuelle Bibliothek? (own translation) -

4. Microsoft Encarta Enzyklopädie 1998. Entry "virtuell"

5. See: Koster, Martijn: Robot Exclusion Standard - and:

6. Hochschulentwicklung durch neue Medien - Vision 2005. BIG - Bildungswege in der Informationsgesellschaft - Gütersloh: 1999

7. ill. 3: To Publish and Perish. In: Policy Perspectives. March 1998, Volume 7, Number 4

8. Richard Heseltine: The Future of the Subject Resource Gateways. Proceedings of the library Strategy workshop. Bristol: JISC, 1998 Borges

9. Informationsinfrastruktur im Wandel - Herausforderungen für die Hochschulen und ihre Informations- und Kommunikationseinrichtungen. Thesen der gemeinsamen Arbeitsgruppe. – Oldenburg: 1998 (own translation)

10. Eco, Umberto: Die Bibliothek. München: Hanser 1987, S. 15 (own translation)

11. Anne G. Lipow: Serving the remote user: reference service in the digital environment.

12. Borges, S. 50f

13. Arnoud de Kemp: Real oder Virtuell? Erfahrungsbericht über die Entwicklung und Nutzung von elektronischen Bibliotheken aus dem Springer-Verlag. In: Service im Wandel. 27. Arbeits- und Fortbildungstagung der AspB/Sektion5 im DBV.- Karlsruhe: ASpB, 1999. p. 127 (own translation)

14. Borges, S. 55

Index